T0137968

Power Systems

For further volumes:
http://www.springer.com/series/4622

Filipe Faria da Silva · Claus Leth Bak

Electromagnetic Transients in Power Cables

 Springer

Filipe Faria da Silva
Department of Energy Technology
Aalborg University
Aalborg
Denmark

Claus Leth Bak
Institute of Energy Technology
Aalborg University
Aalborg
Denmark

Additional material to this book can be downloaded from http://extras.springer.com/

ISSN 1612-1287   ISSN 1860-4676 (electronic)
ISBN 978-1-4471-6222-3   ISBN 978-1-4471-5236-1 (eBook)
DOI 10.1007/978-1-4471-5236-1
Springer London Heidelberg New York Dordrecht

PSCAD^{TM}/EMTDC^{TM} from Manitoba HVDC Research Centre, a division of Manitoba Hydro International Ltd.

© Springer-Verlag London 2013
Softcover re-print of the Hardcover 1st edition 2013
This work is subject to copyright. All rights are reserved by the Publisher, whether the whole or part of the material is concerned, specifically the rights of translation, reprinting, reuse of illustrations, recitation, broadcasting, reproduction on microfilms or in any other physical way, and transmission or information storage and retrieval, electronic adaptation, computer software, or by similar or dissimilar methodology now known or hereafter developed. Exempted from this legal reservation are brief excerpts in connection with reviews or scholarly analysis or material supplied specifically for the purpose of being entered and executed on a computer system, for exclusive use by the purchaser of the work. Duplication of this publication or parts thereof is permitted only under the provisions of the Copyright Law of the Publisher's location, in its current version, and permission for use must always be obtained from Springer. Permissions for use may be obtained through RightsLink at the Copyright Clearance Center. Violations are liable to prosecution under the respective Copyright Law. The use of general descriptive names, registered names, trademarks, service marks, etc. in this publication does not imply, even in the absence of a specific statement, that such names are exempt from the relevant protective laws and regulations and therefore free for general use.
While the advice and information in this book are believed to be true and accurate at the date of publication, neither the authors nor the editors nor the publisher can accept any legal responsibility for any errors or omissions that may be made. The publisher makes no warranty, express or implied, with respect to the material contained herein.

Printed on acid-free paper

Springer is part of Springer Science+Business Media (www.springer.com)

Preface

For more than a century, overhead lines have been the most commonly used technology for transmitting electrical energy at all voltage levels, especially on the highest levels. However, in recent years, an increase in both the number and length of HVAC cables in the transmission networks of different countries like Denmark, Japan or United Kingdom has been observed. At the same time, the construction of offshore wind farms, which are typically connected to the shore through HVAC cables, increased exponentially.

As the number of HVAC cables increased, the interest in the study of electromagnetic phenomena associated to their operation, among them electromagnetic transients, increased as well. Transient phenomena have been studied since the beginning of power systems, at first using only analytical approaches, which limited studies to more basic phenomena; but as computational tools became more powerful, the analyses started to expand for the more complex phenomena.

Being old phenomena, electromagnetic transients are covered in many publications, and classic books such as the 40-year-old Greenwood's "Electric Transients in Power Systems" are still used to this day. However, the majority of publications tend to ignore HVAC cables, which is understandable as the use of long HVAC cables was not very common until recent years.

This book proposes to address some of the transient phenomena that may occur when operating power networks with HVAC cables. The book is written as a textbook and it tries to give comprehensive explanations of the different phenomena and focus on describing different scenarios. It is the authors' opinion that this approach allows for a better understanding of the physical principles and for readers to adapt their analyses accordingly when handling different cases concerning HVAC cables.

An important topic that is not covered in this book is measurements protocols/methods. The protocols used when performing measurements on a cable depend on what is to be measured, the available equipment and accessibility. Readers interested in the topic are referred to search for this information in Ph.D. theses and scientific papers.

However, the book is not only intended for students . It can also be used by engineers who work in this area and need to understand the challenges/problems they are facing or who need to learn how to prepare their simulation models.

Chapter 1, "Components Description", describes the several layers of a cable as well as their function. It also shows how to calculate the different electric parameters of a cable, resistance, inductance, capacitance, and how to use those values to calculate the positive-sequence and zero-sequence impedance, including how to adapt the datasheet values for more accurate calculations.

The chapter introduces the different bonding configurations typically used for the screens of cables (both-ends bonding and cross-boding) and also presents methods that can be used to estimate the maximum current of a cable for different types of soils, i.e. thermal calculations.

The end of the chapter introduces the shunt reactor, which is an important element in cable-based network as it consumes locally the reactive power generated by the cables.

Chapter 2, "Simple Switching Transients", reviews the principles of the Laplace transform and uses it to study simple switching transients, RC–RL–RLC loads for both AC and DC sources.

In other words, the chapter demonstrates how analytical analysis of simple systems can be made. These principles will be used in later chapters in the study of more complex scenarios.

Chapter 3, "Travelling Waves and Modal Domain", reviews the Telegraph equations and how to calculate the loop and series impedance matrices as well as the shunt admittance matrix of a cable in function of the frequency.

The chapter also introduces the different modes of a cable, how to calculate their impedance and velocity, as well as their frequency dependence. The knowledge of modal theory is of utmost importance when working in transient in cables. It is true that in many cases, software is used to run simulations, and the reader may be tempted to think that only those designing the software need to know how to use modal theory. However, several phenomena require at least a minimum knowledge of the topic and for that reason, the book provides a thorough explanation of the subject.

Chapter 3 also studies the frequency spectrum of a cable for different bonding configurations.

Chapter 4, "Transient Phenomena", describes several electromagnetic phenomena that may occur in HVAC cables. The chapter starts by explaining the energisation of a single cable for both-ends bonding and cross-bonding, showing the waveforms for different scenarios and demonstrating how the modal theory can be used to explain the transient waveforms; after this, other phenomena such as the energisation of cables in parallel, zero-missing, transient recovery voltages and restrikes are addressed.

Hybrid cable-OHLs are also considered in this chapter, and it is demonstrated how an overvoltage may be very high for some configurations as well as the influence of the bonding configuration in the magnitude of the overvoltage.

The interaction between a cable which is highly capacitive and a transformer which is highly inductive is also analysed and some possible resonance scenarios are explained, as well as ferroresonance.

To finish the chapter, we study short-circuits in cables, which can be rather different from short-circuit in OHLs, because of the current returning in the screen. The screen can also be bonded on different configurations, influencing both the magnitude of the short-circuit current and of the transient recovery voltage.

Chapter 5, "System Modelling and Harmonics", starts by proposing a method that can be used when deciding how much of the network to model when doing a simulation of an energisation/restrike together with the possible limitations of the method.

The chapter continues by analysing the frequency-spectrums of cable-based networks which have lower resonance frequencies than usual because of the larger capacitance of the cables. At the same time, a technique that may help save time when plotting the frequency spectrum of a network is proposed.

The chapter ends by proposing a systematic method that can be used when doing the insulation co-ordination study for a line, as well as the modelling requirements, both modelling depth and modelling detail of the equipment, for the study of the different types of transients followed by a step-by-step generic example.

Acknowledgments

This book would not be possible without the contribution of several people whose names do not appear on the cover, but still deserve acknowledgment.

Much of the work presented here was developed using Filipe Faria da Silva's Ph.D. Thesis as initial draft. The thesis was made in collaboration with the Danish TSO (Energinet.dk), which supported all the associated expenses and provided first-hand access to all the necessary data. As a result, the authors would like to personally thank Wojciech Wiechowski and Per Balle Holst for co-supervising the project, as well as the entire staff of the Planning and Transmission sections.

The authors would also like to acknowledge the support from the Manitoba HVDC Research Centre for allowing the free distribution of several case-examples made in PSCAD and help with all the software-related problems, no matter how strange the questions were.

The contribution of Christian Flytkjær Jensen for the short-circuit section must be recognised, together with a big thank you for having read that section twice, providing very good hints and suggestions both times.

Both authors were member of the CIGRE WG C4.502, which studied systems with long HVAC cables. The good discussions and work developed by the WG strongly contributed to increase the authors' knowledge of the topic. For that reason, the authors would like to thank all the members for having indirectly contributed to this book.

A word of thanks to Springer editorial staff, especially Grace Quinn, is also deserved, due to all their support.

Finally, Filipe and Claus would like to thank their respective girlfriends, Aida and Tine. Aida for all the patience she had during the long writing process and Tine for having performed a thorough proofreading of the entire book.

Contents

Chapter 1
Components Description

1.1 Cable

The invention of cable technology can be dated as far back as 1830, and the first underground cable installation 50 years later in 1880, in Berlin. The reason for this temporal hiatus was the need to find a dielectric material capable of withstanding the conductor's heat and the strong electric field. This was only achieved with Ferranti's invention in 1880 of a multi-layer dielectric using lapped paper tapes, which was able to fulfil the requirements. This technology was later improved by Emanueli in 1917, by impregnating the paper dielectric with low-viscosity insulating oil under permanent pressure. This improved the cable's thermal stability and made it possible for the first time to use cables for voltages higher than 100 kV.

The next big step forward in power cable technology occurred in the 1960 s with the introduction of cross-linked polyethylene (XLPE) as dielectric, which allowed for higher operational temperatures (around 90 °C). Another common cable technology is high pressure gas-filled (HPGF) pipe-type cable, which typically uses SF_6. These cables have relativity short lengths (<3,000 m) and are not as common as XLPE cables. In this technology, the conductor consists of a rigid aluminium tube inside of an aluminium pipe, the space between the conductor and the pipe being filled with pressurised SF_6 gas. In order to increase the electric strength of the gas, the gas is used under a pressure of 3–5 bar.

The basic design of an HV/EHV cable remained unchanged during the last century. The main cable components are: the conductor, the insulation, and the metallic screen (see Fig. 1.1).

Conductor
The conductor is typically made of copper or aluminium, and its main function is to carry the electrical current. The size of the conductor is determined by the current flowing through it, the cross section being proportional to the current, and also by the dielectric used, at present mainly XLPE.

F. F. da Silva and C. L. Bak, *Electromagnetic Transients in Power Cables*,
Power Systems, DOI: 10.1007/978-1-4471-5236-1_1,
© Springer-Verlag London 2013

The copper has the advantage of having higher tensile strength and a lower resistivity than aluminium, resulting in a smaller cross section for the same current. However, the aluminium has a lower density than copper, making an aluminium cable approximately three times lighter than a copper cable with the same ampacity, which is a big advantageous when installing very long cables, because of the transport. In the end, the decision between copper or aluminium is based on the price.

Insulation

The insulation is one of the most important cable components. Its function is to ensure that there is not an electrical connection between the conductor and the cable's screen and to maintain an uniformly divergent field. Therefore, the insulation must be capable of withstanding the cable's electric field for steady-state and transient conditions.

Historically the insulation was fluid-impregnated paper insulation, but this material become less popular in recent years for environmental reasons, being substituted by extruded polymers. The most common polymer is the XLPE, which is used in the majority of HV cables produced at the present. The XLPE insulation has also the advantage of operating at a maximum steady-state temperature of 90 °C against the 60 °C of paper insulation, allowing transmitting more power for the same conductor.

Screen

The metallic screen/sheath's main function is to nullify the electric field outside of the cable, provide a return path for the charging current and to conduct fault currents to the earth. Other advantages of using metal screens are mechanical protection of the cable against accidental contact and minimisation of proximity effect. The size of metallic screen depends on the value of the zero-sequence short-circuit current that must be drained by the cable.

Metallic screens of HV cables are often, but not always, made of copper wires together with an aluminium foil.

Semiconductive layers

The use of semiconductive layers between the conductor and the insulation, and between the insulation and the metallic screen aims to ensure a cylindrical electric field and to avoid the formation of gaps or voids between the conductor/screen/metal sheath, preventing the occurrence of partial discharges.

In order to avoid the formation of gaps is necessary that material used in the semiconductive layers to be similar to the one of the insulation. As a result, modified extrudable polymers with a large conductivity are used for manufacturing the semiconductive layers of cables with XLPE insulation.

Jackets or armours are added when the cable is installed in tougher environments for mechanical protection. They provide also insulation between the screen and the earth.

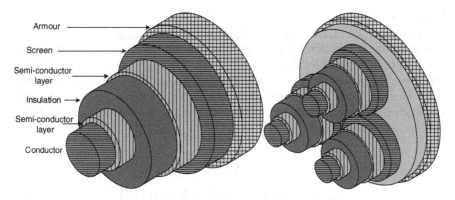

Armour
Screen
Semi-conductor layer
Insulation
Semi-conductor layer
Conductor

Fig. 1.1 A typical XLPE single-core cable and a typical pipe-type cable

Pipe-type cable

A pipe-type cable is a cable where all three-cores are inside a common armour or pipe and each core is similar to the single-core cable just described. Some sources distinguish between having the three cables inside an armour or a pipe and use different names. In the former, fillers are used to keep the cables in place and they are normally equidistant to the central point. In the latter, the cables are inside a pipe insulated with a gas or oil; in this case the cables can be at the bottom of the pipe. In this book, we use the term pipe-type cable for both and distinguish between them if necessary.

In some cases, mainly at distribution lever, there is a common screen to all three phases, instead of one for each phase, meaning that the insulation is also common to all three phases. In this case there is capacitive coupling between the phases.

Figure 1.1 shows a typical single-core XLPE cable and a typical pipe-type cable.

1.1.1 Electrical Parameters

Before we advance to the transient studies we should take some time learning how to do the calculations for steady-state conditions. As an example, there is no point at the planning stage, when only load flows are studied, to use the complex equations necessary for the analysis of transient phenomena.

The backbone of the theory presented in the next pages is the classic theory that many of the readers already learned for OHLs. There are differences in the calculation of the electric parameters that are explained next, but pi-models can typically continued to be used and the long-line corrections continue to be applied on the same way.

Resistance

The DC resistance of a conductor depends on the material, the size of the cross section and the temperature. The resistance is calculated using (1.1), where ρ is the electrical resistivity and S the cross-section area. Yet, as the electrical resistivity is normally given for 20 °C it is necessary to adjust the value at the temperature (1.2), where α_T is the temperature coefficient. Table 1.1 shows the resistivity and temperature coefficient for copper and aluminium the two materials used to build the conductor of a cable.

$$R_{DC} = \frac{\rho}{S} \left[\Omega.\text{m}^{-1}\right] \tag{1.1}$$

$$R_{DC}(T) = R_{20°}(1 + \alpha_T(T - 20)) \left[\Omega.\text{m}^{-1}\right] \tag{1.2}$$

The previous expression is used to calculate the resistance of a conductor for DC conditions. However, the resistance of a conductor is larger for AC currents than DC currents, because of *skin effect* and *proximity effect*.

Skin effect is a result of the current not to be uniformly distributed over the conductor for AC currents. Instead, the current tends to concentrate on the surface of the conductor as the frequency increases due to electromagnetic induction. The concentration of the current on the surface of the conductor is equivalent to a reduction of the cross-section area, resulting in an increase of the resistance as the frequency increases.

Proximity effect happens when two or more conductors are nearby and there is an AC current in at least one of them. This AC current influences the current distribution in the other conductor(s) by inducing eddy currents that oppose the original current. Similar to skin effect, the proximity effect increases the resistance and it is frequency dependent.

As a result, it is necessary to include these two factors in the calculation of the resistance, which is given by (1.3), where y_s and y_p are the skin effect factor and proximity effect factor, respectively. For pipe-type cables the equation changes to (1.4).

$$R = R_{DC}(1 + y_s + y_p) \left[\Omega.\text{m}^{-1}\right] \tag{1.3}$$

$$R = R_{DC}(1 + 1.5(y_s + y_p)) \left[\Omega.\text{m}^{-1}\right] \tag{1.4}$$

The skin and proximity effects factors are calculated using the empirical formulas (1.5) and (1.7) respectively.

Table 1.1 Resistivity and temperature coefficient of copper and aluminium [11]

	Resistivity (Ω.m)	Temperature Coeff. (per K at 20 °C)
Copper	1.724×10^{-8}	3.93×10^{-3}
Aluminium	2.286×10^{-8}	4.03×10^{-3}

Table 1.2 Skin and proximity effect—experimental values for the coefficients k_s and k_p

Type of conductor	k_s	k_p
Copper		
Round, solid	1	1
Round, stranded	1	1
Aluminium		
Round, solid	1	1
Round, stranded	1	0.8

$$y_s = \frac{x_s^4}{192 + 0.8x_s^4} \tag{1.5}$$

$$x_s^2 = \frac{8\pi f}{R_1}10^{-7}k_s \tag{1.6}$$

$$y_p = \frac{x_p^4}{192 + 0.8x_p^4}\left(\frac{d_c}{s}\right)^2 \cdot \left[0.312\left(\frac{d_c}{s}\right)^2 + \frac{1.18}{\frac{x_p^4}{192 + 0.8x_p^4} + 0.27}\right] \tag{1.7}$$

$$x_p^2 = \frac{8\pi f}{R_1}10^{-7}k_p \tag{1.8}$$

The coefficients k_s and k_p are experimental values that depend on the conductor's material and shape. Table 1.2 shows the values that should be used according to the most recent standard [11].[1]

Solid conductors are typically not used in cables with large cross sections, because of high skin and proximity effects. Instead, there are used segmented or stranded conductors that minimise both skin effect and proximity effect, allowing larger cross sections.

As a result, the cross section of the conductor is smaller than πR_1, because of the empty spaces between the stranded/segmented wires of the conductor and it is necessary to correct the resistivity value (1.9). The calculated value is then used to calculate the resistance of the conductor

$$\rho = \rho_{Sol}\frac{\pi R_1}{S} \; [\Omega.\text{m}] \tag{1.9}$$

Capacitance
The capacitance of a coaxial cable is given by the classic expression (1.10). Where, ε is the permittivity of the insulation, R_2 the radius over the insulation including the semi-conductors layers and R_1 the radius over the conductor.

---

[1] The table shows only some of the values, for the full table consult the standard [11]. The values presented are for the future issue of the standard, which is not yet approved. Thus, the final version may have slightly different values.

$$C = \frac{2\pi\varepsilon}{\ln\left(\frac{R_2}{R_1}\right)} \; [\text{F.m}^{-1}] \tag{1.10}$$

Semi-conductive layers are installed between the conductor and the insulation, and between the insulation and the screen. Thus, the capacitance between the conductor and the screen is equivalent to a series coupling of the capacitance in the former semi-conductive layer, in the insulation and in the latter semi-conductive layer.

Therefore, the semi-conductive layers are considered as part of the insulation and the permittivity is corrected according to (1.11). Where, ε_{Ins} is the insulation permittivity that is typically between 2.3 and 2.5, b the outer-radius of the insulation, and a the inner-radius of the insulation.

$$\varepsilon = \varepsilon_{\text{Ins}} \frac{\ln\left(\frac{R_2}{R_1}\right)}{\ln\left(\frac{b}{a}\right)} \; [\text{F.m}^{-1}] \tag{1.11}$$

Inductance

The reactance of the conductor is typically calculated using the classic expression (1.12). Where μ is the permeability of the conductor, typically the one of the vacuum, D_e is the depth penetration of the earth and GMR the geometric mean radius of the conductor, normally equal to $R_1 e^{-1/4}$.

$$L = \frac{\mu}{2\pi} \ln\left(\frac{D_e}{GMR}\right) \; [H.\text{m}^{-1}] \tag{1.12}$$

The depth of the earth is calculated using (1.13). Where, ρ_{earth} is the resistivity of the earth.

$$D_e = 659\sqrt{\frac{\rho_{\text{earth}}}{f}} \; [\Omega.\text{m}] \tag{1.13}$$

1.1.2 Sequence Impedances

The positive-sequence impedance depends on the type of bonding used in the cable (see Sect. 1.2), whereas the zero-sequence impedance is independent of the bonding, except for single-point bonding, which is not typically used for HV cables.

The positive-sequence impedance of a cross-bonded cable is calculated by (1.14), whereas for a cable bonded at both ends is calculated using (1.15). The zero-sequence impedance is calculated using (1.16) for both bonding configurations.

$$Z^+_{\text{Cross}} = (Z_{\text{Self}} - Z_M) \tag{1.14}$$

$$Z^+_{\text{Both-ends}} = (Z_{\text{Self}} - Z_M) - \frac{(Z_{M,S} - Z_M)^2}{Z_{\text{Self},S} - Z_M} \tag{1.15}$$

$$Z^0 = Z_{\text{Self}} + 2Z_M - \frac{(Z_{M,S} + 2Z_M)^2}{Z_{\text{Self},S} + 2Z_M} \tag{1.16}$$

where:

- Z_{Self} is the self-impedance of the conductor, calculated according to (1.17)
- $Z_{\text{Self},S}$ is the self-impedance of the screen, calculated according to (1.18)
- Z_M is the mutual-impedance between cables, calculated according to (1.19), where s is the distance between phases.
- $Z_{M,S}$ is the mutual-impedance between the conductor and screen, calculated according to (1.20)

$$Z_{\text{Self}} = R_{50\,\text{Hz}} + R_e + jX_L \tag{1.17}$$

$$Z_{\text{Self},S} = R_S + R_e + jX_S \tag{1.18}$$

$$Z_M = R_e + j\frac{\omega\mu}{2\pi}\ln\left(\frac{D_e}{s}\right) \tag{1.19}$$

$$Z_{M,S} = R_e + j\frac{\omega\mu}{2\pi}\ln\left(\frac{D_e}{R_2}\right) \tag{1.20}$$

If the cable is installed in flat formation it is necessary to calculate the mutual inductance (1.19) between adjacent phases (Z_{M_i}) and the outer phases (Z_{M_o}), being the total mutual inductance the average as given by (1.21):

$$Z_M = \frac{2Z_{M\_i} + Z_{M\_o}}{3} \tag{1.21}$$

It is important to refer that the frequency corrections made for the resistance of the conductor are not usually made for the screen, because of its low thickness.

1.1.3 Example

We are now going to calculate the positive-sequence and zero-sequence impedances of a 1,200 mm^2 cable using the equations previously described for a cable installed in a trefoil formation operating at 90 °C and 50 Hz. Table 1.3 shows the cable's data.

The core of the cable is made of compact stranded aluminium wires. It is therefore necessary to correct the resistivity of the core (1.22). Using the corrected resistivity, it calculated the resistance of the conductor (1.23).

Table 1.3 Cable data

Layer	Thickness (mm)	Material
Conductor	41.5[a]	Aluminium, round, compacted
Conductor screen	1.5	Semi-conductive PE
Insulation	17	Dry cured XLPE
Insulation screen	1	Semi-conductive PE
Longitudinal water barrier	0.6	Swelling tape
Copper wire screen	95[b]	Copper
Longitudinal water barrier	0.6	Swelling tape
Radial water barrier	0.2	Aluminium laminate
Outer cover	4	High-density PE
Complete cable	95[a]	–

[a] Diameter
[b] Cross-section

$$\rho = \rho_{Sol}\frac{\pi R_1}{S} \Leftrightarrow \rho = 2.826 \cdot 10^{-8}\frac{\pi \cdot 20.75^2}{1200} \Leftrightarrow \rho = 3.186 \cdot 10^{-8} \, [\Omega.m] \quad (1.22)$$

$$R_{DC} = \frac{\rho}{S} \Leftrightarrow R_{DC} = \frac{3.186 \cdot 10^{-8}}{\pi(20.75 \cdot 10^{-3})^2} \Leftrightarrow R_{DC} = 23.55 \cdot 10^{-6} \, [\Omega.m^{-1}] \quad (1.23)$$

The resistance is then corrected to 90 °C (1.24).

$$R_{DC}90° = 23.55 \cdot 10^{-6}\left(1 + 4.03 \cdot 10^{-3}(90 - 20)\right)$$
$$\Leftrightarrow R_{DC}90° = 30.19 \cdot 10^{-6} \, [\Omega.m^{-1}] \quad (1.24)$$

At last we have the skin and proximity effects. The frequency of the calculations is 50 Hz and skin and proximity effects should both be small. The equations confirm the expected and the skin effect factor is 0.133 (1.26), whereas the proximity effect factor is 0.018 (1.27).

$$x_s^2 = \frac{8\pi f}{R_{DC}}10^{-7}k_s \Leftrightarrow x_s^2 = \frac{8\pi 50}{23.55 \cdot 10^{-6}}10^{-7} \cdot 1 \Leftrightarrow x_s^2 = 5.336 \quad (1.25)$$

$$y_s = \frac{x_s^4}{192 + 0.8x_s^4} \Leftrightarrow y_s = 0.133 \quad (1.26)$$

$$y_p = \frac{x_p^4}{192 + 0.8x_p^4}\left(\frac{d_c}{s}\right)^2 \cdot \left[0.312\left(\frac{d_c}{s}\right)^2 + \frac{1.18}{\frac{x_p^4}{192+0.8x_p^4} + 0.27}\right] \Leftrightarrow$$

$$\Leftrightarrow y_p = \frac{(5.336)^2}{192 + 0.8(5.336)^2}\left(\frac{40.5}{2 \cdot 95}\right)^2 \cdot \left[0.312\left(\frac{40.5}{2 \cdot 95}\right)^2 + \frac{1.18}{\frac{(5.336)^2}{192+0.8(5.336)^2} + 0.27}\right]$$

$$\Leftrightarrow y_p = 0.018$$

$$(1.27)$$

Thus, the resistance of the conductor is equal to (1.28):

$$R_{50Hz} = 30.19 \cdot 10^{-6}(1 + 0.133 + 0.018) \Leftrightarrow R_{50Hz} = 34.75 \cdot 10^{-6} \left[\Omega.m^{-1}\right]$$
(1.28)

The reactance of the conductor is calculated as shown in (1.30).

$$D_e = 659\sqrt{\frac{100}{50}} \Leftrightarrow D_e = 932 \, [m]$$
(1.29)

$$X_L = 2\pi 50 \frac{\mu}{2\pi} \ln\left(\frac{932}{20.75 \cdot 10^{-3} e^{-\frac{1}{4}}}\right) \Leftrightarrow X_L = 6.9 \cdot 10^{-4} \left[\Omega.m^{-1}\right]$$
(1.30)

Having calculated all the parameters associated with the conductor, there are calculated parameters associated with the screen. The resistance of the screen is calculated in (1.31), (1.32) and the reactance in (1.33).

$$R_S = \frac{\rho}{S} \Leftrightarrow R_S = \frac{1.724 \cdot 10^{-8}}{\pi\left((42.76 \cdot 10^{-3})^2 - (40.85 \cdot 10^{-3})^2\right)}$$
(1.31)
$$\Leftrightarrow R_S = 3.44 \cdot 10^{-5} \left[\Omega.m^{-1}\right]$$

$$R_{S,90°} = R_S(1 + \alpha_T(T - 20)) \Leftrightarrow R_{S,90°} = 3.44 \cdot 10^{-5}\left(1 + 3.98 \cdot 10^{-3}(90 - 20)\right)$$
$$\Leftrightarrow R_{S,90°} = 4.40 \cdot 10^{-5} \left[\Omega.m^{-1}\right]$$
(1.32)

$$X_S = \frac{\omega\mu}{2\pi} \ln\left(\frac{D_e}{\frac{R_2+R_3}{2}}\right) \Leftrightarrow X_S = 50\mu \ln\left(\frac{932}{41.085 \cdot 10^{-3}}\right)$$
(1.33)
$$\Leftrightarrow X_S = 6.30 \cdot 10^{-4} \left[\Omega.m^{-1}\right]$$

The only parameter missing for the calculation of the self-impedance is the ground impedance, which can be calculated according to the classic Carson's formula (1.34).

The self-impedance is then given by the summation of the conductor's resistance with the conductor's reactance and the ground impedance (1.35). The self-impedance of the screen is similar but the resistance and reactance of the conductor are replaced by the ones of the screen (1.36).

$$R_e = 9.869 \cdot 10^{-7} f \left[\Omega.m^{-1}\right]$$
(1.34)

$$Z_{Self} = R_{50 Hz} + R_e + jX_L \Leftrightarrow Z_{Self} = 34.75 \cdot 10^{-6} + 49.35 \cdot 10^{-6} + j6.9 \cdot 10^{-4} \Leftrightarrow$$
$$\Leftrightarrow Z_{Self} = 84.10 \cdot 10^{-6} + j6.9 \cdot 10^{-4} \left[\Omega.m^{-1}\right]$$
(1.35)

$$Z_{Self,S} = R_S + R_e + jX_S \Leftrightarrow Z_{Self,S} = 44.0 \cdot 10^{-6} + 49.35 \cdot 10^{-6} + j6.3 \cdot 10^{-4}$$
$$\Leftrightarrow Z_{Self,S} = 93.35 \cdot 10^{-6} + j6.3 \cdot 10^{-4} \left[\Omega.m^{-1}\right]$$
(1.36)

The mutual impedance between phases is calculated in (1.37) using the classic formula, whereas the mutual impedance between conductor and screen is calculated in (1.38).

$$Z_M = R_e + j\frac{\omega\mu}{2\pi}\ln\left(\frac{D_e}{s}\right) \Leftrightarrow Z_M = 49.35 \cdot 10^{-6} + j5.34 \cdot 10^{-4}\left[\Omega.m^{-1}\right]$$

(1.37)

$$Z_{M,S} = R_e + jX_S \Leftrightarrow Z_{M,S} = 49.35 \cdot 10^{-6} + j6.30 \cdot 10^{-4}\left[\Omega.m^{-1}\right] \qquad (1.38)$$

The positive-sequence impedance for a cross-bonded cable is calculated in (1.39) and for a cable bonded in both ends in (1.40).

$$Z^+ = Z_{\text{Self}} - Z_M \Leftrightarrow Z^+ = 34.75 \cdot 10^{-6} + j1.56 \cdot 10^{-4}\left[\Omega.m^{-1}\right] \qquad (1.39)$$

$$Z^+ = (Z_{\text{Self}} - Z_M) - \frac{\left(Z_{M,S} - Z_M\right)^2}{Z_{\text{Self},S} - Z_M}$$

$$\Leftrightarrow Z^+ = 36.25 \cdot 10^{-6} + j1.64 \cdot 10^{-4}\left[\Omega.m^{-1}\right]$$

(1.40)

The zero-sequence impedance for both bonding configurations is given by (1.41).

$$Z^0 = Z_{\text{Self}} + 2Z_M - \frac{\left(Z_{M,S} + 2Z_M\right)^2}{Z_{\text{Self},S} + 2Z_M}$$

$$\Leftrightarrow Z^0 = 74.32 \cdot 10^{-6} + j2.47 \cdot 10^{-4}\left[\Omega.m^{-1}\right]$$

(1.41)

Finally, we calculate the capacitance of the cable, but before doing it, it is necessary to adjust the permittivity (1.42) and to adjust the radius used in the calculation of the capacitance value (1.43).

$$\varepsilon = \varepsilon_{\text{Ins}}\frac{\ln\left(\frac{R_2}{R_1}\right)}{\ln\left(\frac{b}{a}\right)} \Leftrightarrow \varepsilon = 2.5\frac{\ln\left(\frac{40.85}{20.75}\right)}{\ln\left(\frac{39.50}{22.25}\right)} \Leftrightarrow \varepsilon' = 2.95 \qquad (1.42)$$

$$C = \frac{2\pi\varepsilon}{\ln\left(\frac{R_2}{R_1}\right)} \Leftrightarrow C = \frac{2\pi 2.95 \cdot 8.85 \cdot 10^{-12}}{\ln\left(\frac{40.85}{20.75}\right)} \Leftrightarrow C = 2.42 \cdot 10^{-10}\left[F.m^{-1}\right] \;(1.43)$$

1.1.4 Other Losses

Dielectric Losses
Dielectric losses are often not considered in the calculation of the losses, particularly at distribution levels. However, at transmission levels these losses may be rather high, as they are proportional to the square of the voltage, and must be addressed.

Dielectric losses are voltage-dependent losses, which depend also on the capacitance of the dielectric, i.e. insulation, the frequency and the loss factor (tan δ).[2]

$$W_d = \omega \cdot C \cdot U_0^2 \cdot \tan(\delta) \tag{1.44}$$

Circulating Current Losses
See Sect. 1.2 Bonding Techniques.

1.2 Bonding Techniques

In steady-state conditions the current circulating in the core of a cable induces a voltage in the screen of the same cable. If a closed loop is formed there is a current circulating in the screen, increasing the system losses. Thus, it is important to keep the current in the screen as low as possible.

The screens of three-phase cable are normally installed in one of three bonding configurations:

• Single-end bonding—Grounding of the screens at one end only;
• Both-ends bonding—Grounding of the screens at both ends;
• Cross-bonding—Grounding of the screens at both ends, with transposition of the screens in between ends.

The single-end bonding has the advantage of avoid currents circulating in the screen, as there is no close loop. As effective as this technique is to reduce the circulating current losses, it has the big disadvantageous of leading to an increase of the screen voltage. One end of the screen is grounded at zero potential, whereas the other end of the screen has a voltage, whose magnitude is proportional to the length of the cable. As a result, this technique is limited to short cables, typically less than 3 km, for which the voltage in the screen is still within tolerable values.

The both-ends bonding to the screen reduces the voltage in the screen, which is now virtually zero in both ends, to very low values. However, there is a close path for the current in the screen and the losses are larger. In some situations, these losses can be as high as the conductor losses.

The cross bonding is a middle term between the two previous bonding techniques. It eliminates neither the circulating currents nor the induced voltage, but it reduces both to low values. The cable is divided into three sections, called minor sections. The screens are transposed between minor sections and grounded at every third minor section, forming a major section. The minor sections should have a similar length in order to keep the system as balanced as possible and cable may have as many major sections as necessary. Figure 1.2 shows a diagram of a major section of a cross-bonded cable.

---

[2] 0.001 for a HV XLPE-type cable.

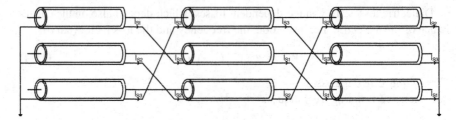

Fig. 1.2 Example of a major section of a cross-bonded cable

The transposition of the screens assures that each screen is exposed to the magnetic field generated by each phase. Assuming a balance system, i.e. same magnitude and 120° phase difference between phases, installed in trefoil configuration and that each minor section had exactly the same length, the induced voltages would cancel and the circulating current would be null. However, in reality it is not possible to have these perfect conditions, as neither the minor sections have precisely the same length nor the currents in the conductor have the same magnitude, and there is a circulating current. Moreover, HVAC cables are often installed in flat formation, resulting in an unbalanced mutual-coupling between phases.

Due to all the reasons previously stated, HV cables are normally installed using the cross-bonding technique. At the distribution level it is normal to find also cables installed using the both-ends bonding technique. For practical reasons the length of land-cable that can be carried in a drum is usually limited to a maximum of 3 km, limiting the typical length of a minor-section to less than 3 km.[3]

The most notable exception of cables not being installed in cross-bonded formations at HV are submarine cables, which have to be installed using both-ends bonding, since it is unpractical to transpose the screens on the sea.

1.3 Cable Thermal Behaviour

The power transmitted by a cable is limited by the maximum operating temperature of the cable, typically 90 °C for modern cables. Consequently, the layout conditions and the installation path chosen to the cable are of outmost importance and have direct influence on the ampacity of the cable.

In this section, we will learn how to estimate the thermal resistances of a cable and use them to estimate the cable ampacity. The term thermal resistance is used because the thermal resistances can be combined into a thermal equivalent circuit using the *Ohm's law of heat conduction*, which works on the same principles of the Ohm's law for electric circuits.

---

[3] Assuming transmission voltage levels.

In other words, the thermal resistances can be manipulated as electric resistances using the same rules. The conductor of the cable is the heat source and can be seen as equivalent to a voltage/current source in an electric circuit, whereas the atmosphere is the heat sink and is equivalent to ground in an electric circuit. The temperature drop between the conductor and the air is equivalent to the voltage drop between the source and the ground and the heat flow is equivalent to the current. The thermal resistances are associated with temperature drops and are equivalent to voltage drops in an electric circuit.

The non-conductive layers between the conductor and the air are the thermal resistances and are divided into internal and external thermal resistances. The conductive layers of a cable, i.e. the metallic parts, are good heat conductors and are considered as isothermal.

Internal thermal resistances are typically three[4]:

- T_1: Thermal resistance between conductor and screen
- T_2: Thermal resistance between screen and armour/ground
- T_3: Thermal resistance for outer covering of the cable, between the armour and the ground (only present if the cable has an armour)

The external thermal resistance (T_4) is formed by the soil surrounding the cable. The external thermal resistance is the most important of all the thermal resistance and it can be often responsible for differences in the cable ampacity in the order of hundreds of amps.

Figure 1.3 shows the distribution of the thermal resistance in a single-core cable and an artificial electric circuit that it is equivalent to.

Before explaining how to calculate the different thermal resistances it is important to point out that in this book we only study single-core cables and pipe-type cables, which are the most typical configurations of high-voltages. However, there are other configurations that are possible and common at distribution levels that are not analysed in this chapter, as are cables installed in ducts or tunnels. For those interested in these cases it is advised to consult the standard [12].

1.3.1 Internal Thermal Resistances

In the next pages are shown the formulas used for the calculation of the different thermal resistances.[5]

The thermal resistance between conductor and screen (T_1) is calculation according to (1.45). Where, ρ_T is the thermal resistivity of the insulation [K.m/W] and R_1 are R_2 are the radius as shown in Fig. 1.3.

---

[4] The semi-conductive layers are considered in the insulation for effects of calculation.

[5] For consistency, the notation used for some of the variables is slightly different from the one used in the standard.

Fig. 1.3 Thermal equivalent circuit for a single-core cable

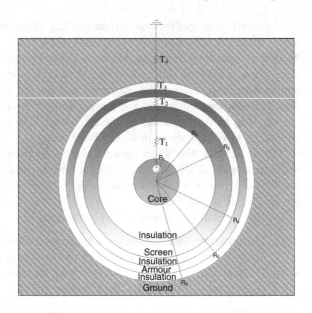

$$T_1 = \frac{\rho_T}{2\pi}\ln\left(1 + \frac{2(R_2 - R_1)}{2R_1}\right) \tag{1.45}$$

The thermal resistance between screen and armour (T_2) and between the armour and ground (T_3) are calculated using similar formulas, shown in (1.46) and (1.47), respectively. One should remember that the thermal resistance T_3 only exists if the cable has armour.

$$T_2 = \frac{\rho_T}{2\pi}\ln\left(1 + \frac{2(R_4 - R_3)}{2R_4}\right) \tag{1.46}$$

$$T_3 = \frac{\rho_T}{2\pi}\ln\left(1 + \frac{2(R_6 - R_5)}{2R_6}\right) \tag{1.47}$$

The previous formulas are for single-core cables, for pipe-type cables it is necessary to change some of the parameters. Assuming the traditional pipe-type configuration core-screen-pipe we have:

- T_1 remains unchanged as the pipe does not affect the insulation between the conductor and the screen;
- T_2 is now the thermal resistance between the screen and the pipe. There is one pipe for three screens and thus the formula needs to be changed. The thermal resistance is given by (1.48), where is a geometric factor, whose graph is available in [10].
- T_3 is now the thermal resistance between the pipe and the ground. It is calculated using the same formula, but it is necessary to adjust the radiuses.

$$T_2 = \frac{\rho_T}{6\pi}\overline{G} \tag{1.48}$$

1.3.2 External Thermal Resistances

The external thermal resistance (T_4) is more difficult to calculate than the internal thermal resistances and it is strongly dependent on the soil and installation conditions.

We start by analysing a single-core cable buried underground without any other cables in its vicinity (1.49).

$$T_4 = \frac{\rho_T}{2\pi}\ln\left(u + \sqrt{u^2 - 1}\right) \tag{1.49}$$

where, ρ_T is the soil resistivity (strongly dependent on the soil type)

- u is equal to $2\,h/(2R_6)$
- h is the distance from the surface of the ground to the cable axis
- R_6 is the radius of the external layer of the cable (having Fig. 1.3 as reference)

The formula (1.49) is valid for a cable standing alone, but the most common situation it to have three cables, or more, together. In this situation it used the method of images (if the cables are not touching).

We assume that the cables are of the same type and have the same load.[6] In this situation, the formula (1.49) is slightly changed to consider the influence of the other cables (1.50). The new term F_B is calculated using the method of images as shown in (1.51) for one phase of a three-phase single-core cable.

$$T_4 = \frac{\rho_T}{2\pi}\ln\left(\left(u + \sqrt{u^2 - 1}\right)F_B\right) \tag{1.50}$$

$$F_B = \left(\frac{d'_{21}}{d_{21}}\right)\cdot\left(\frac{d'_{23}}{d_{23}}\right) \tag{1.51}$$

The formula can be further expanded for N phases by using the method of images (1.52).

$$T_4 = \frac{\rho_T}{2\pi}\ln\left(\left(u + \sqrt{u^2 - 1}\right)\left[\left(\frac{d'_{p1}}{d_{p1}}\right)\cdot\left(\frac{d'_{p2}}{d_{p2}}\right)\cdots\left(\frac{d'_{pN}}{d_{pN}}\right)\right]\right) \tag{1.52}$$

---

[6] If the cables have different loads the value of T_4 is calculated for each cable individually are the temperature calculation is changed (please refer to the standard [12] for this case).

If the cables are touching and in a symmetric configuration, i.e. a trefoil formation, the thermal resistance is calculated using (1.53) for metallic sheathed cables and (1.54) for non-metallic sheathed cables.

$$T_4 = \frac{1.5\rho_T}{\pi} (\ln(2u) - 0.630) \tag{1.53}$$

$$T_4 = \frac{\rho_T}{2\pi} (\ln(2u) + 2\ln(u)) \tag{1.54}$$

For the purpose of calculating the external thermal resistance a pipe-type cable is similar to a single-core cable standing alone and the calculation is made using (1.49). For several pipe-type cables installed together there are used the expressions previously shown.

1.3.3 Ampacity Calculation

The ampacity is one of the most important parameters when choosing the cable as it defines the maximum steady-state current. If no current flows, the cable is at ambient temperature. As the current increases the temperature in the cable also increases in a relation given by (1.55).

$$\Delta\theta = (I^2R + \tfrac{1}{2}W_d) \cdot T_1 + [I^2R \cdot (1 + \lambda_1) + W_d] \cdot n \cdot T_2$$
$$+ [I^2R \cdot (1 + \lambda_1 + \lambda_2) + W_d] \cdot n \cdot (T_3 + T_4) \tag{1.55}$$

where: $\Delta\theta$ is the conductor temperature which raise above the ambient temperature [K or °C]

- I is the current flowing in the conductor [A]
- R is the AC resistance per unit of length at maximum operating temperature [$\Omega.\text{m}^{-1}$]
- W_d is the dielectric losses per unit of length for the main insulation [W.m^{-1}]
- T_{1-4} are the thermal resistances [K.m.W^{-1}]
- n is the number of load-carrying conductors in the cable
- λ_1 is the ratio of losses in the metal screen to total losses in all conductors in the cable
- λ_2 is the ratio of losses in the armouring to total losses in all conductors in the cable.

The relation between the temperature and the current is far from being linear and it is dependent of many factors. The equation also shows that as expected it is important to keep the electrical resistance and the thermal resistances as low as possible.

For our analysis we are going to break the equation into three parts.

By means of Joule losses heat is generated when a current flows in the conductor. At the same time there is a voltage difference in the dielectric of the cable, i.e. the insulation, and there are present dielectric losses, which do also release heat. The dielectric losses are uniformly distributed along the insulation, but for numerical reasons these losses are considered to be in an infinitesimal thin cylinder located at the centre of the insulation. Therefore, only half of the heat released by the dielectric losses in the insulation is associated with T_1.

The precise extension of the temperature increase is obtained by multiplying these two types of losses by the thermal resistance T_1, which corresponds to the main insulation. In other words, the lower the thermal resistivity of the insulation is, the lower is the temperature increase.

Next, we focus our attention in the thermal resistance T_2 between screen and armour. Joule losses continue to be present, but it is necessary to add the Joule losses in the conductor to the Joule losses in the screen (λ_1). The dielectric losses also continue to be presented, but are now full and not half, as the thermal resistance T_2 is over the insulation. The number of conductors depends on the cable structure and it is 1 for single-core cables and 3 for three-core cables sharing the same screen.

The reasoning applied to the thermal resistance T_2 is applied to the thermal resistances T_3 and T_4 and are also added to the Joule losses in the armour (λ_2).

The equation (1.55) can be manipulated and the permissible current rating is obtained (1.56). The value of $\Delta\theta$ varies from country to country as it depends on the temperature of the soil, but it is usually between 65 and 80 C for XLPE-type cables.

$$
I = \left[\frac{\Delta\theta - W_d \cdot [0.5 \cdot T_1 + n \cdot (T_2 + T_3 + T_4)]}{R \cdot T_1 + n \cdot R \cdot (1 + \lambda_1) \cdot T_2 + n \cdot R \cdot (1 + \lambda_1 + \lambda_2) \cdot (T_3 + T_4)} \right]^{\frac{1}{2}}
$$

$$(1.56)$$

1.3.4 Example

The objective is to calculate the ampacity of the cable used in the previous example (Sect 1.1.3). We start by calculating the thermal resistances.

The thermal resistivities are as follows:

- Insulation (XLPE and High Density PE): 3.5 K.m.W^{-1}
- Soil: 0.9 K.m.W^{-1} (typical safe value)

These values are used to calculate the thermal resistances. T_1 and T_2 are calculated directly using respectively (1.57) and (1.58) m whereas T_3 does not exist as the cable has no armour.

$$T_1 = \frac{\rho_T}{2\pi}\ln\left(1 + \frac{2(R_2 - R_1)}{2R_1}\right) \Leftrightarrow T_1 = \frac{3.5}{2\pi}\ln\left(1 + \frac{2(40.85 - 20.75)}{2 \cdot 20.75}\right)$$
$$\Leftrightarrow T_1 = 0.377\,\mathrm{m^2K \cdot W^{-1}} \tag{1.57}$$

$$T_2 = \frac{\rho_T}{2\pi}\ln\left(1 + \frac{2(R_4 - R_3)}{2R_4}\right) \Leftrightarrow T_2 = \frac{3.5}{2\pi}\ln\left(1 + \frac{2(47.5 - 42.76)}{2 \cdot 47.5}\right) \tag{1.58}$$
$$\Leftrightarrow T_2 = 0.053\mathrm{m^2K \cdot W^{-1}}$$

$$T_3 = 0\,\mathrm{m^2K \cdot W^{-1}} \tag{1.59}$$

The thermal resistance T_4 is calculated using the expression for trefoil cables with a metallic sheath (the aluminium foil) (1.60). The distance from the surface of the ground to the cable axis is calculated considering that the top cable is installed at a depth of 1.1 m.

$$T_4 = \frac{1.5\rho_T}{\pi}(\ln(2u) - 0.630) \Leftrightarrow T_4 = \frac{1.5 \cdot 0.9}{\pi}\left(\ln\left(2\frac{2h}{2R_4}\right) - 0.630\right)$$
$$\Leftrightarrow T_4 = \frac{1.5 \cdot 0.9}{\pi}\left(\ln\left(2\frac{2\left(\frac{1.1+1.18185}{2}\right)}{2 \cdot 47.5 \times 10^{-3}}\right) - 0.630\right) \Leftrightarrow T_4 = 1.393\ \mathrm{m^2K \cdot W^{-1}} \tag{1.60}$$

It is also necessary to calculate the dielectric losses (1.61) and the ratio of losses in the screen of the cable (1.62).

$$W_d = \omega \cdot C \cdot U_0^2 \cdot \tan(\delta) \Leftrightarrow W_d = 2\pi50 \cdot 2.42 \cdot 10^{-10} \cdot \left(165 \cdot 10^3\right)^2 0.001$$
$$\Leftrightarrow W_d = 207\ \mathrm{Wm^{-1}} \tag{1.61}$$

$$\lambda_1 = RS, 90^\circ\ R_{50\mathrm{Hz}}\frac{1}{1 + \left(\frac{R_S}{X_S}\right)^2} \Leftrightarrow \lambda_1 = \frac{4.40 \cdot 10^{-5}}{34.75 \cdot 10^{-6}}\frac{1}{1 + \left(\frac{4.40\cdot10^{-5}}{6.30\cdot10^{-4}}\right)^2} \tag{1.62}$$
$$\Leftrightarrow \lambda_1 = 1.26$$

In possession of all the information we can now calculate the maximum steady-state current (1.63).

$$I = \left[\frac{75 - 2.07 \cdot [0.5 \cdot 0.377 + (0.053 + 1.393)]}{34.75 \cdot 10^{-6} \cdot 0.377 + 34.75 \cdot 10^{-6} \cdot (1 + 1.26) \cdot 0.053 + 34.75 \cdot 10^{-6} \cdot (1 + 1.26) \cdot 1.393}\right]^{\frac{1}{2}}$$
$$\Leftrightarrow I = 752\ \mathrm{A} \tag{1.63}$$

Besides the temperature of the soil, which is behind human control, the other factor with a strong influence in the maximum current is the thermal resistance T_4. As an example, if the same cable was installed close to the surface at a depth of 0.5 m instead of 1.1 m, the thermal resistance would reduce to 1.072 and the maximum steady-state current would increase to 844 A.

It is also possible to improve the thermal conductivity of the soil around the cable by using special types of soil. The same cable installed in a soil with a thermal resistivity of 0.6 K.m.W^{-1} would have at a maximum steady-state current of 897 A.

1.4 Shunt Reactor

The capacitance of a cable is 10–20 times higher than the capacitance of an equivalent OHL. Such high capacitance results in high generation of reactive power by the cable, which has to be consumed in order to ensure that the voltage does not increase and that the ampacity of the cable is not reduced. The consumption of this reactive power is achieved by using shunt reactors, which are defined as an *inductive reactance whose purpose is to draw inductive current from the system* [10].

In order to fully understand the need for shunt reactors we have first to understand why the voltage increases in a cable, contrarily to what is usual for OHLs, in a phenomenon typically known as *Ferranti Effect*.

Ferranti Effect occurs when energising an unloaded or lightly loaded line in such way that the capacitive current of the line is greater than the load. As a result, Ferranti Effect is more pronounced in cables than in OHLs, because of their large capacitance.

As an example, for an unloaded line the voltage in the receiving end is calculated, for a lossless line, using a pi-model by (1.64), where V_1 and V_2 are the sending and receiving end voltages, respectively, L the line inductance, C the line capacitance and l the line length.

$$V_1 - V_2 = I \cdot j\omega Ll \Leftrightarrow V_1 - V_2 = \left(V_2 j\omega \frac{Cl}{2} \right) \cdot j\omega Ll \Leftrightarrow V_2 = V_1 \frac{1}{1 - \frac{\omega^2 LCl^2}{2}}$$

(1.64)

Observing (1.64), it is obvious that the voltage increase is more accentuated in cables than OHLs, because of the larger capacitance of the cables. The voltage increase also depends on the square of the cable length, making the phenomenon worse for long lines. In reality, if an OHL is long enough it will also experience a voltage increase from the sending end to the receiving end.

According to (1.64) the voltage would always increase even for short OHLs, but that is because the line was considered lossless. In the real world the resistance of the line would reduce the voltage in the receiving end and it would be smaller than in the sending end for a typical OHL or a short cable.[7]

---

[7] At the end of this chapter it is proposed an exercise for the calculation of the equation for a line with losses. The solution is available online, together with some plots showing possible scenarios.

Now that we understood why the voltage increases, it is time to understand the importance of the shunt reactors. A power system has strict limits for the voltage magnitude, which has to be maintained inside a limited interval. Additionally, all the capacitive current generated by the cable reduces the ability of the cable to carry active power. If not consumed the reactive power propagates into the neighbour lines reducing the also their ampacities.

For both reasons are installed shunt reactors when laying long HVAC cables.

The location of the shunt reactor(s) is also of most importance both for steady-state and transients.[8]

Figure 1.4 shows the loading of an open cable for different reactive power compensation schemes. The best location for the shunt reactor is in the middle of the cable as the current flows from the shunt reactor into both directions. As an example, the installation of the shunt reactor in one of the cable ends would result in two times the amount of reactive current flowing in the cable, and the installation of one shunt reactor in each of the cable ends would have the same effect as the installation of one shunt reactor in the middle of the cable. In a similar way, the reactive current flowing in the cable is reduced when using more shunt reactors distributed along the cable.

We will see that many transient phenomena are closely related and dependent on the shunt reactors that are installed together with the cables. To some of the phenomena it is advantageous to have the shunt reactor, whereas in other it would be beneficial not to have it. Thus, it is important to understand how they operate their specific characteristics.

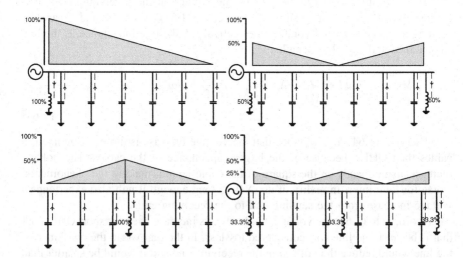

Fig. 1.4 Loading of an open cable for different reactive power compensation schemes

---

[8] Because of the interaction between the shunt reactor and the cable during the transient (more on Chap. 4).

A shunt reactor is like a big inductor and in many aspects comparable to a transformer. Some of the electromagnetic phenomena associated with shunt reactors are:

- Saturation
- Mutual coupling
- Magnetisation losses
- Copper Losses
- Additional losses (stray losses, skin effect, etc...)

The modelling of shunt reactors is typically done by an inductor in series with a resistor and it is not common to consider the effects just enumerated, except for phenomena with very-high frequencies.

This simple model is very often accurate enough, but it will be demonstrated later in the book that for some phenomena it is necessary also to model the mutual inductance between the shunt reactor phases and the shunt reactor saturation in order to achieve an accurate simulation.

The principles of mutual coupling should be familiar to most of the readers. In a nutshell, the coils of a shunt reactor are typically mounted in the same iron core. Therefore, when current flows in one phase, a magnetic flux flows into all limbs of the shunt reactor, as shown in Fig. 1.5. The flux links with the two other coils inducing a voltage in the corresponding phases.

Table 1.4 shows the voltage induced in each phase when applied a voltage of 98.15 kV_{RMS} for a typical 80 MVA shunt reactor. It is seen that the induced voltage is rather small when compared with the applied, but it is high enough to influence the disconnection of a cable and shunt reactor together.

Fig. 1.5 Flux path for a voltage applied to phase A of a 5-limbs shunt reactor

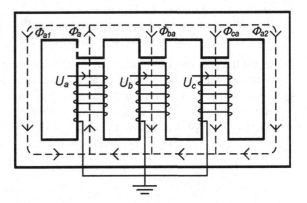

Table 1.4 Induced voltages due to mutual coupling

Phase	Applied voltage (kV)	Induced voltage (kV)		
		A	B	C
A	98.15	–	1.18 (0.51 %)	0.39 (0.17 %)
B	98.15	1.02 (0.44 %)	–	1.09 (0.47 %)
C	98.15	0.30 (0.13 %)	1.09 (0.47 %)	–

1.5 Exercises

1) Calculate the positive-sequence and zero-sequence impedance for the following cable with a cross section of 800 mm^2, bonded at both ends, installed 1 m deep in flat formation with 1 m between conductors in a soil with a grounding resistance of 100 Ω/m: (A: $Z^+ = 75.59 \cdot 10^{-6} + j1.17 \cdot 10^{-4}$ [$\Omega.m^{-1}$]; $Z^0 = 77.49 \cdot 10^{-6} + j1.08 \cdot 10^{-4}$ [$\Omega.m^{-1}$])

Layer	Thickness (mm)	Material
Conductor	33.7[a]	copper, round, compacted
Conductor screen	1.5	Semi-conductive PE
Insulation	19	Dry cured XLPE
Insulation screen	1	Semi-conductive PE
Aluminium tape	600[b]	Aluminium
Outer cover	4	High-density PE
Complete cable	89.5[a]	–

[a] Diameter
[b] Cross-section

2) Calculate the ampacity for the cable of the previous exercise when installed in a soil with a thermal resistivity equal to 0.9 K.m/W and a temperature of 25 °C. The thermal resistivity of the insulations is 3.5 K.m/W. (A: I = 845.4 A)

3) Repeat the previous exercise, but assuming that the cable is installed in a trefoil formation with the top conductor at a 1 m deep. (A: I = 659.5 A)

4) Calculate the expression for voltage increase due to Ferranti Effect for an unloaded line using a pi-model (similar to Sect. 1.4, but for a line with losses).
   (A: $\|V_2\| = \|V_1\| \dfrac{1}{\sqrt{1+\left(\frac{\omega^2 LCl^2}{2}\right)^2 - \omega^2 LCl^2 + \left(\frac{\omega RCl^2}{2}\right)^2}}$)

References and Further Reading

1. Moore GF (2007) Electric cables handbook, 3rd edn. Blackwell Science, Oxford
2. Benato R, Paolucci A (2010) EHV AC undergrounding electric power. Springer, London

3. Peschke E, von Olshausen R (1999) Cable systems for high and extra-high voltage—development, manufacture, testing, installation and operation of cables and their accessories. Publicis MCD Werbeagentur GmbH, Munich
4. Popović Zoya, Popović BD (2000) Introductory electromagnetic. Prentice Hall, New Jersey
5. Anders GJ (1997) Rating of electric power cables. IEEE press power engineering series, New York
6. Tziouvaras DA (2006) Protection of high-voltage AC cables, 59th annual conference for protective relay engineers
7. Cigre WG C4.502 (2013) Power system technical performance issues related to the application of long HVAC cables, CIGRE, Paris
8. Cigre WG B1.30 (2013) Cable systems electrical characteristics, Cigre, Paris
9. Cigre WG B1.19 (2004) General guidelines for the integration of a new underground cable system in the networks, Cigre, Paris
10. IEEE Standard C57.21 (1990) IEEE standard requirements, terminology, and test code for shunt reactors rated over 500 kVA, IEEE, New York
11. EC Standard 60287-1-1 (2006) Electric cables—calculation of the current rating—Part 1-1: Current rating equations (100 % load factor) and calculation of losses—general, IEC, Geneva
12. IEC Standard 60287-2-1 (2006) Electric cables—Calculation of the current rating—Part 2-1: Thermal resistance—calculation of thermal resistance—general, IEC, Geneva

Chapter 2
Simple Switching Transients

2.1 Laplace Transform

The Laplace transformation is among the most useful tools used by electrical engineers. The purpose of this book is not to provide a mathematical treatment of the Laplace transform. Instead, the next pages focus on the practical application of the method and its use on different types of circuits.

The Laplace transformation is a method that can be used for solving ordinary differential equations by reducing a differential equation to an algebraic equation which can then be solved by the more common and easy algebraic operations. Moreover, the Laplace transformation is a linear operation, i.e., $L(af(t) + bg(t)) = aL(f(t)) + bL(g(t))$ and it can be applied to piecewise continuous functions, meaning that the function may have finite "jumps".

The Laplace transform of a function $f(t)$ is defined by (2.1). However, it is normally unnecessary to solve the equation as tables of transforms exist for the more common expressions. Table 2.1 shows some of the main transformations used by electrical engineers.

$$F(s) = L(f(t)) = \int_0^\infty e^{-st} f(t) \, dt \tag{2.1}$$

One question that may arise is: can Laplace transform applied to any function. The theory states that the transform can be applied for any $s > \gamma$ as long as the condition (2.2) is fulfilled for all $t \geq 0$ and for some constant M and γ [1]. In other words, the transform exists if $e^{-st} f(t)$ goes to zero when $t \to \infty$, something which is true for any physical system, since that *for any real physical stimulus there will be a real physical response* [2].

$$|f(t)| \leq Me^{\gamma t} \tag{2.2}$$

F. F. da Silva and C. L. Bak, *Electromagnetic Transients in Power Cables*,
Power Systems, DOI: 10.1007/978-1-4471-5236-1_2,
© Springer-Verlag London 2013

Table 2.1 Some Laplace transforms of the more common functions

f(t)	F(s)		f(t)	F(s)		f(t)	F(s)
I	1	$\frac{1}{s}$	IV	e^{at}	$\frac{1}{s-a}$	VII	$e^{-a\|t\|}$	$\frac{2a}{a^2-s^2}$
II	t	$\frac{1}{s^2}$	V	$1-e^{-at}$	$\frac{a}{s(s+a)}$	VIII	$\cos(\omega t)$	$\frac{s}{s^2+\omega^2}$
III	t^n	$\frac{n!}{s^{n+1}}$	VI	$\frac{t^n}{n!}e^{-at}$	$\frac{1}{(s+a)^{n+1}}$	IX	$\sin(\omega t)$	$\frac{\omega}{s^2+\omega^2}$

As mentioned the Laplace transform is normally used to solve differential equations. These equations contain a combination of derivative and integration functions and in the following, we will see how to work with these functions.

The Laplace transform of a differentiation corresponds to the multiplication of the transform $F(s)$ by the complex Laplace variable s, whereas the Laplace transform of an integration corresponds to a division. In other words, the differential equation becomes a polynomial equation.

Equation (2.3) shows how to apply the Laplace transform to a first order derivate. The variable s multiplies the transformation and the initial value of the function f is subtracted. It is important to keep in mind that the solution to a differential equation has two parts, the *general* solution and the *particular* solution. The general solution is a characteristic of the system being studied, and it is independent of the system condition [left term in (2.3)], whereas the particular solution depends on the system condition, normally the system initial conditions [right term in (2.3)].

$$\mathrm{L}\left(f'(t)\right) = s\mathrm{L}(f(t)) - f(0) \tag{2.3}$$

The same method can be applied to higher order derivate functions, with minor changes. Equation (2.4) shows how to apply the Laplace transform to a second order derivate. The deduction of the formula is rather straightforward and shown in (2.5).

$$\mathrm{L}\left(f''(t)\right) = s^2\mathrm{L}(f(t)) - sf(0) - f'(0) \tag{2.4}$$

$$\begin{aligned}\mathrm{L}(f''(t)) &= s\mathrm{L}(f'(t)) - f'(0)\\ &= s(s\mathrm{L}(f) - f(0)) - f'(0)\\ &= s^2\mathrm{L}(f(t)) - sf(0) - f'(0)\end{aligned} \tag{2.5}$$

The reasoning applied in (2.5) can be applied to any higher order derivate function and the following general expression is obtained (2.6).

$$\mathrm{L}(f^n(t)) = s^n\mathrm{L}(f(t)) - s^{n-1}f(0) - s^{n-2}f'(0) - \cdots - f^{n-1}(0) \tag{2.6}$$

The integration is the opposite operation of derivation, and thus, there is a division instead of a multiplication. Equation (2.7) shows how to apply the Laplace transform to an integral defined between 0 and a random time t.

$$\mathrm{L}\left(\int_0^t f(t)\mathrm{d}t\right) = \frac{1}{s}\mathrm{L}(f(t)) \tag{2.7}$$

Do not worry if it seems complicated. The use of the Laplace transformation of the different proprieties will become clearer via the examples shown on the next pages.

2.2 Switching of RL Circuits (or Shunt Reactors)

The first circuit that we are going to analyse is a basic RL circuit. An RL circuit is a first-order circuit and due to its simplicity, it is a good choice for an introduction to the world of transients. The circuit also resembles the behaviour of a shunt reactor which is an important piece of equipment in any high voltage cable-based network. Figure 2.1 shows the single-line of an RL circuit consisting of a resistance and inductance in series connected to a voltage source through a switch (CB).

2.2.1 DC Source

The simplest example is the energisation of the RL load by a DC voltage source. Such a system is mathematically described by (2.8). The application of the Laplace transform to (2.8) leads to (2.9).

$$V = RI + L\frac{\mathrm{d}I}{\mathrm{d}t} \tag{2.8}$$

$$V = RI(s) + L(sI(s) - I(0)) \tag{2.9}$$

The circuit is considered as being discharged previous to the switching and thus $I(0) = 0$, because of the inductor, and the solution to the equation is the one shown in (2.10).

Fig. 2.1 Switching of an RL load

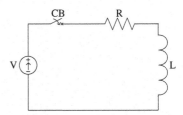

$$\frac{V}{s} = RI(s) + sLI(s) \Leftrightarrow I(s) = \frac{V}{s(R+sL)} \Leftrightarrow I(s) = \frac{V}{L}\frac{1}{s(R/L+s)} \quad (2.10)$$

Applying the inverse Laplace transform (V in Table 2.1) to (2.10) allows to obtain the time domain expression of the current (2.11).

$$I(t) = L^{-1}(I(s)) \Leftrightarrow I(t) = L^{-1}\left(\frac{V}{L}\frac{L}{R}\frac{R/L}{s(R/L+s)}\right)$$

$$\Leftrightarrow I(t) = \frac{V}{R}\left(1 - e^{-\frac{R}{L}t}\right) \quad (2.11)$$

We can see that the current starts at zero and increases up to V/R with a time constant (τ) equal L/R and with the behaviour of an inverse exponential decay function. It is common to consider the current in steady-state after approximately 5τ.

It is important to keep in mind that the current cannot change instantaneously due to the conservation of the moment in the magnetic flux associated to the inductor. In other words, the magnetic flux has to be continuous, i.e., an instant change in the current would require an infinite voltage which is obviously impossible in a real system.

Example:
Figure 2.2 shows the current during the switching of an RL load for the following parameters: $V = 100$ V, $R = 1\ \Omega$ and $L = 0.1$ H.

We know from (2.11) that the time constant of the circuit is $L/R = 0.1$ s and the steady-state value is equal to $V/R = 100$ A. Figure 2.2 confirms the results showing a current with the shape of an inverse exponential decaying function with an approximate steady-state value of 100 A after 0.5 s (5×0.1).

Fig. 2.2 Current in the load during the first 0.5 s when using a DC voltage source

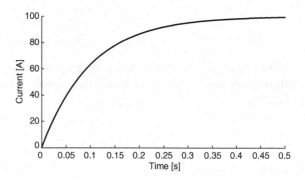

2.2.2 AC Source

The previous example was for a DC circuit, however, the majority of the power systems are AC. The change of the voltage source from DC to AC leads to radical changes in both the equation and the waveform.

The first step is to solve the equation describing the circuit (2.12) and apply the Laplace transform (2.13). For simplification, it is considered that the circuit is switched at zero voltage and that it was unloaded prior to the energisation, $I(0) = 0$ A.

$$V_P\sin(\omega t) = RI + L\frac{dI}{dt} \tag{2.12}$$

$$V_P\frac{\omega}{s^2 + \omega^2} = I(R + sL) - I(0) \tag{2.13}$$

$$I = V_P\frac{\omega}{s^2 + \omega^2}\frac{1}{R + sL} \Leftrightarrow I = V_P\frac{\omega}{L}\frac{1}{s^2 + \omega^2}\frac{1}{s + \frac{R}{L}} \tag{2.14}$$

The resolution of (2.14) requires the use of the partial fraction method as done in (2.15) and (2.16), where $a = R/L$.

$$\frac{1}{s^2 + \omega^2}\frac{1}{s + a} = \frac{As + B}{s^2 + \omega^2} + \frac{C}{s + a} \tag{2.15}$$

$$\begin{cases} s^2 : A + C = 0 \\ s^1 : Aa + B = 0 \\ s^0 : Ba + C\omega^2 = 1 \end{cases} \tag{2.16}$$

By solving (2.16), (2.17) is obtained. Using (2.17) in (2.15) and later in (2.14), (2.18) is obtained.

$$A = \frac{-1}{a^2 + \omega^2} \wedge B = \frac{a}{a^2 + \omega^2} \wedge C = \frac{1}{a^2 + \omega^2} \tag{2.17}$$

$$I = V_P\frac{\omega}{L}\frac{1}{a^2 + \omega^2}\left(\frac{-s + a}{\omega^2 + s^2} + \frac{1}{s + a}\right)$$
$$\Leftrightarrow I = V_P\frac{\omega}{L}\frac{1}{a^2 + \omega^2}\left(\frac{-s}{s^2 + \omega^2} + \frac{a}{s^2 + \omega^2} + \frac{1}{s + a}\right) \tag{2.18}$$

The inverse Laplace transform can be applied to (2.18) leading to (2.19). Notice how the linearity of the Laplace transform act as an advantage in this case.

$$I(t) = V_P \frac{\omega}{L} \frac{1}{a^2 + \omega^2} \left(-\cos(\omega t) + \frac{a}{\omega} \sin(\omega t) + e^{-at} \right)$$

$$\Leftrightarrow I(t) = \frac{V_P}{L} \frac{1}{\sqrt{a^2 + \omega^2}} \left(-\frac{\omega}{\sqrt{a^2 + \omega^2}} \cos(\omega t) + \frac{a}{\sqrt{a^2 + \omega^2}} \sin(\omega t) + \frac{\omega}{\sqrt{a^2 + \omega^2}} e^{-at} \right)$$

$$(2.19)$$

It is known that the power factor [$\cos(\phi)$] of an RL load is given by (2.20) which is equivalent to (2.21).

$$\cos(\phi) = \frac{R}{\sqrt{R^2 + (\omega L)^2}} \tag{2.20}$$

$$\cos(\phi) = \frac{R}{L\sqrt{\left(\frac{R}{L}\right)^2 + \omega^2}} \Leftrightarrow \cos(\phi) = \frac{a}{\sqrt{a^2 + \omega^2}} \tag{2.21}$$

Similar relations can be obtained for $\sin(\phi)$ (2.22) and $\tan(\phi)$ (2.23).

$$\sin^2(\phi) + \cos^2(\phi) = 1 \Leftrightarrow \sin(\phi) = \sqrt{1 - \frac{a^2}{a^2 + \omega^2}} \Leftrightarrow \sin(\phi) = \frac{\omega}{a^2 + \omega^2}$$

$$(2.22)$$

$$\tan(\phi) = \frac{\sin(\phi)}{\cos(\phi)} \Leftrightarrow \tan(\phi) = \frac{\omega}{a} \tag{2.23}$$

Replacing (2.21) and (2.22) in (2.19), (2.24) is obtained.

$$I(t) = \frac{V_P}{L} \frac{1}{\sqrt{a^2 + \omega^2}} \left(-\sin(\phi)\cos(\omega t) + \cos(\phi)\sin(\omega t) + \sin(\phi)e^{-at} \right) \tag{2.24}$$

Equation (2.24) can be further simplified using the trigonometric relation (2.25), leading to (2.26).

$$\cos(\phi)\sin(\omega t) - \sin(\phi)\cos(\omega t) = \sin(\omega t - \phi) \tag{2.25}$$

$$I(t) = \frac{V_P}{L} \frac{1}{\sqrt{a^2 + \omega^2}} \left(\sin(\omega t - \phi) + \sin(\phi)e^{-at} \right)$$

$$\Leftrightarrow I(t) = \frac{V_P}{L} \frac{1}{\sqrt{a^2 + \omega^2}} \left(\sin\left(\omega t - \tan^{-1}\left(\frac{\omega}{a}\right)\right) + \sin\left(\tan^{-1}\left(\frac{\omega}{a}\right)\right)e^{-at} \right)$$

$$\Leftrightarrow I(t) = \frac{V_P}{\sqrt{R^2 + (\omega L)^2}} \left(\sin\left(\omega t - \tan^{-1}\left(\omega\frac{L}{R}\right)\right) - \sin\left(-\tan^{-1}\left(\omega\frac{L}{R}\right)\right)e^{-\frac{R}{L}t} \right)$$

$$(2.26)$$

The equation were developed assuming that the circuit was switched on for zero voltage. However, the circuit can be energised at any chosen instant. A process

Fig. 2.3 Switching angle in
a sinusoidal wave

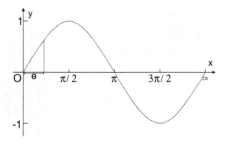

similar to the one just done in the last pages can be done, which would result in
(2.27), where θ is the switching angle (Fig. 2.3).

$$I(t) = \frac{V_P}{\sqrt{R^2 + (\omega L)^2}} \left(\sin\left(\omega t + \theta - \tan^{-1}\left(\omega \frac{L}{R}\right)\right) - \sin\left(\theta - \tan^{-1}\left(\omega \frac{L}{R}\right)\right) e^{-\frac{R}{L}t} \right)$$

$$(2.27)$$

As previously stated, a shunt reactor is basically an RL circuit and it is
described by (2.27). However, the high inductance of a shunt reactor which is
typically hundreds of times larger than the resistance allows simplification of
(2.27) to the more compact (2.28).

$$I(t) \simeq \frac{V_P}{\sqrt{R^2 + (\omega L)^2}} \left(\sin\left(\omega t + \theta - \frac{\pi}{2}\right) - \sin\left(\theta - \frac{\pi}{2}\right) e^{-\frac{R}{L}t} \right) \qquad (2.28)$$

We obtained an equation describing a general RL circuit, but we have not yet
analysed and explained it. The current in (2.28) is the summation of two parts:

- The steady-state component (also known as forced regime): $\sin\left(\omega t + \theta - \frac{\pi}{2}\right)$;
- The transient component (also known as homogeneous regime): $\sin\left(\theta - \frac{\pi}{2}\right) e^{-\frac{R}{L}t}$.

The steady-state component of the current is basically a sinusoidal wave
oscillating at power frequency with a phase difference of approximately 90° to the
voltage, and it is independent of the switching instant or conditions.

The transient component is a decaying DC current whose amplitude depends on
the initial conditions, i.e., switching instant and the energy stored in the inductance. The energy stored is typically zero ($I(0) = 0$ A), but the switching instant
can be any, depending on the application and type of CB.

To better understand the high importance of the switching instant, let's see two
examples.

Energisation at peak voltage:
The energisation of the RL load for peak voltage is equivalent to having
$\theta = \pm 90°$. As a consequence, the transient component is zero and the current is
only the steady-state component.

Fig. 2.4 Current in the load during the first 0.5 s when using an AC voltage source and energising at peak voltage

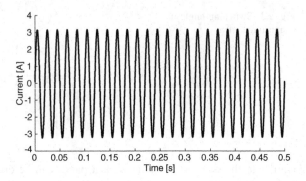

Figure 2.4 shows the energisation of the RL circuit for peak voltage where only the steady-state component is present.

Energisation at zero voltage:

The energisation of the RL load for zero voltage is equivalent to having $\theta = 0°$. The steady-state component is the same for the previous example, as it would be for any example that we can come up with for this circuit. However, the transient component is radically different.

The transient component is a result of the need of maintaining the continuity of the current at the inductor. The steady-state current component has a phase difference of almost 90° to the voltage. Due to energy conservation, the current in the inductor must be continuous; therefore, the transient component has an initial value equal to that of the steady-state current in the connection moment with an opposite sign. Thus, if the RL load is energised when the voltage is zero, the DC component will be at its maximum with a value which is in theory equal to the peak value of the steady-state component.

Figure 2.5 shows the energisation of the RL circuit for zero voltage. Notice the decaying DC component whose initial amplitude is equal to the peak value of the steady-state component, ~ 3.2 A. The decaying rate of the DC component depends

Fig. 2.5 Current during the first 0.5 s when using an AC voltage and energising at zero voltage. *Solid line* current in the RL load, *dashed line* transient component

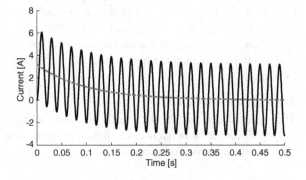

on the time constant R/L, and the smaller the time constant, the longer it takes to damp the transient.

The DC component in the previous example is positive, but it can also be negative. A voltage passes twice by zero during a cycle, and the DC component is maxima for both zeros. However, the signal of the DC component depends on derivate of the voltage. If the voltage is going from negative values to positive values (positive derivate) the DC component has a positive signal; if the voltage is going from positive values to negative values (negative derivate) the DC component has a negative signal.

2.2.3 Summary

In this section, we took contact with our first transient, the energisation of an RL load. We saw how to use the Laplace transform to solve the system and how the switching instant influences the transient. In the case of a RL circuit, the peak current doubles if the circuit is energised at zero-voltage when compared with an energisation at peak voltage. Thus, contrary to what many believe, in this specific case, it would be advantageous to energise at peak voltage instead of zero voltage.

The influence of the switching instant on the transient waveform is common to many transients and it is very often the difference between a smooth transient and a highly undesired transient, as we shall see in Chap. 4.

2.3 Switching of RC Circuits (or Capacitor Banks)

After analysing an RL circuit, we are ready for the next step which is the RC circuit. The RC circuit is also a first order circuit and it resembles the behaviour of a capacitor bank. Figure 2.6 shows the single-line of an RC circuit consisting in a resistance and capacitor in series, connected to a voltage source through a switch.

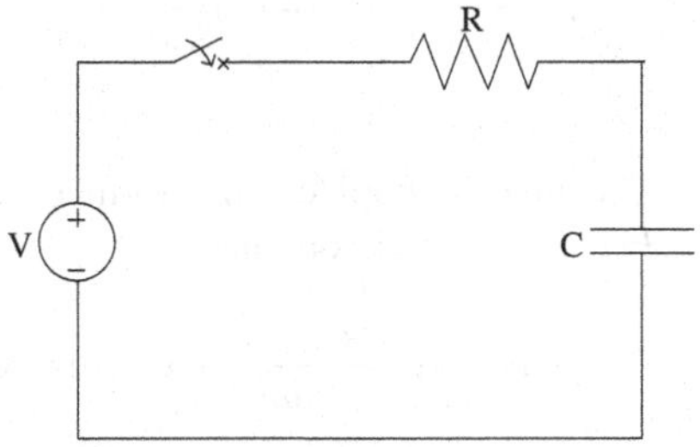

Fig. 2.6 Switching of an RC load

2.3.1 AC Source

An RC load behaves like an open circuit for a DC current, and its transient is identical to the one obtained when the load is connected to an AC source at peak voltage. Thus, we do an analysis for an AC source. Like before, the first thing to do is to write the equation describing the system (2.29) and to apply the Laplace transform (2.30). In order to be able to generalise and become even more acquainted with the Laplace transform, we now consider that the switch can be at any given instant.

$$V_P \sin(\omega t + \theta) = RI + \frac{1}{C} \int I dt \tag{2.29}$$

$$V_P \left(\frac{\omega\cos(\theta)}{s^2 + \omega^2} + \frac{s\sin(\theta)}{s^2 + \omega^2} \right) = RI + \frac{1}{C}\frac{I}{s} + \frac{0}{s}$$

$$\Leftrightarrow I = V_P \frac{saC}{s+a} \left(\frac{\omega\cos(\theta)}{s^2 + \omega^2} + \frac{s\sin(\theta)}{s^2 + \omega^2} \right), \text{ where } a = \frac{1}{RC} \tag{2.30}$$

The easiest way to solve the equation is to divide the second term of (2.31) into two parts (2.32).

$$I = V_P aC \left(\frac{s\omega\cos(\theta)}{(s^2 + \omega^2)(s + a)} + \frac{s^2\sin(\theta)}{(s^2 + \omega^2)(s + a)} \right) \tag{2.31}$$

$$(1) : \frac{s\omega\cos(\theta)}{(s^2+\omega^2)(s+a)}$$
$$(2) : \frac{s^2\sin(\theta)}{(s^2+\omega^2)(s+a)} \tag{2.32}$$

By applying the partial fraction method to (2.32), (2.33) is obtained. Replacing (2.33) in (2.31), (2.34) is obtained.

$$(1) : \frac{\omega\cos(\theta)}{\omega^2+a^2} \left(\frac{sa+\omega^2}{s^2+\omega^2} - \frac{a}{s+a} \right)$$
$$(2) : \frac{\sin(\theta)}{\omega^2+a^2} \left(\frac{s\omega^2-a\omega^2}{s^2+\omega^2} + \frac{a^2}{s+a} \right) \tag{2.33}$$

$$I = V_P \frac{aC}{a^2 + \omega^2} \left[\omega\cos(\theta) \left(\frac{sa + \omega^2}{s^2 + \omega^2} - \frac{a}{s+a} \right) + \sin(\theta) \left(\frac{s\omega^2 - a\omega^2}{s^2 + \omega^2} + \frac{a^2}{s+a} \right) \right] \tag{2.34}$$

The inverse Laplace transform (VI, VIII and IX in Table 2.1) is applied to (2.34) and (2.35) is obtained.

$$I(t) = V_P \frac{aC}{a^2 + \omega^2} [\omega\cos(\theta)(a\cos(\omega t) + \omega\sin(\omega t) - ae^{-at})]$$
$$+ \sin(\theta) (\omega^2\cos(\omega t) - a\omega\sin(\omega t) + a^2 e^{-at}) \tag{2.35}$$

The power factor of an RC load is given by (2.36), which is equivalent to (2.37).

$$\cos(\phi) = \frac{R}{\sqrt{R^2 + \left(\frac{1}{\omega C}\right)^2}} \tag{2.36}$$

$$\cos(\phi) = \frac{1}{\frac{1}{R}\sqrt{R^2 + \left(\frac{1}{\omega C}\right)^2}} \Leftrightarrow \cos(\phi) = \frac{1}{\sqrt{1 + \left(\frac{1}{\omega RC}\right)^2}} \Leftrightarrow \cos(\phi)$$

$$= \frac{1}{\frac{1}{\omega}\sqrt{\omega^2 + \left(\frac{1}{RC}\right)^2}} \Leftrightarrow \cos(\phi) = \frac{\omega}{\sqrt{\omega^2 + a^2}} \tag{2.37}$$

Similar relations can be obtained for $\sin(\phi)$ (2.38) and $\tan(\phi)$ (2.39).

$$\sin^2(\phi) + \cos^2(\phi) = 1 \Leftrightarrow \sin(\phi) = \sqrt{1 - \frac{\omega^2}{a^2 + \omega^2}} \Leftrightarrow \sin(\phi) = \frac{a}{a^2 + \omega^2} \tag{2.38}$$

$$\tan(\phi) = \frac{\sin(\phi)}{\cos(\phi)} \Leftrightarrow \tan(\phi) = \frac{a}{\omega} \tag{2.39}$$

Substituting (2.37)–(2.39), (2.40) is obtained.

$$I(t) = V_P \frac{aC}{\sqrt{a^2 + \omega^2}} \omega\cos(\theta)(\sin(\phi)\cos(\omega t) + \cos(\phi)\sin(\omega t) - \sin(\phi)e^{-at})$$
$$+ V_P \frac{aC}{\sqrt{a^2 + \omega^2}} \omega\sin(\theta)\left(\cos(\phi)\cos(\omega t) - \sin(\phi)\sin(\omega t) + \frac{a}{\omega}\sin(\phi)e^{-at}\right) \tag{2.40}$$

Equation (2.40) can be further simplified using the trigonometric relation (2.42) and (2.42), leading to (2.43).

$$\sin(\phi)\cos(\omega t) + \cos(\phi)\sin(\omega t) = \sin(\omega t + \phi) \tag{2.41}$$

$$\cos(\phi)\cos(\omega t) - \sin(\phi)\sin(\omega t) = \cos(\omega t + \phi) \tag{2.42}$$

$$I(t) = V_P \frac{aC}{\sqrt{a^2 + \omega^2}}$$
$$\left[\omega\cos(\theta)(\sin(\omega t + \phi) - \sin(\phi)e^{-at}) + \omega\sin(\theta)\left(\cos(\phi + \omega t) + \frac{a}{\omega}\sin(\phi)e^{-at}\right)\right] \tag{2.43}$$

Equation (2.43) can be written as (2.44).

$$I(t) = \frac{V_P}{\sqrt{R^2 + \left(\frac{1}{\omega C}\right)^2}}\cos(\theta)\left(\sin\left(\omega t + \tan^{-1}\left(\frac{1}{\omega RC}\right)\right) - \sin\left(\tan^{-1}\left(\frac{1}{\omega RC}\right)\right)e^{-\frac{1}{RC}t}\right) +$$
$$+ \frac{V_P}{\sqrt{R^2 + \left(\frac{1}{\omega C}\right)^2}}\sin(\theta)\left(\cos\left(\omega t + \tan^{-1}\left(\frac{1}{\omega RC}\right)\right) + \frac{1}{\omega RC}\sin\left(\tan^{-1}\left(\frac{1}{\omega RC}\right)\right)e^{-\frac{1}{RC}t}\right)$$

$$(2.44)$$

A capacitance of either a cable or a capacitor bank is in the order of micro-farad. Thus, (2.44) can be simplified into (2.45).

$$I(t) = \frac{V_P}{\sqrt{R^2 + \left(\frac{1}{\omega C}\right)^2}}\left[\cos(\theta)\left(\sin\left(\omega t + \frac{\pi}{2}\right) - e^{-\frac{1}{RC}t}\right) + \sin(\theta)\left(\cos\left(\omega t + \frac{\pi}{2}\right) + \frac{1}{\omega RC}e^{-\frac{1}{RC}t}\right)\right]$$

$$(2.45)$$

Another consequence of having typically a capacitance of micro-farad is the low value of the time constant $\tau = RC$. Meaning that an energisation transient is damped in some micro-seconds; something that must be taken into account when choosing the simulation time-step.

Energisation at zero voltage:

Like before, we are going to separate the analysis of the energisation of the load at zero voltage and the energisation of the load at peak voltage into two cases, starting with the former.

The energisation of the RC load for zero voltage is equivalent to having $\theta = 0°$. Thus, (2.45) can be simplified into (2.46).

$$I(t) = \frac{V_P}{\sqrt{R^2 + \left(\frac{1}{\omega C}\right)^2}}\left(\sin\left(\omega t + \frac{\pi}{2}\right) - e^{-\frac{1}{RC}t}\right) \qquad (2.46)$$

The equation shows us that the current starts at zero and rises while the transient component is damped. It was previously stated that the time constant is typically very high and that the transient component is damped in just some micro-seconds. As a result, the current rises to a value approximately equal to the peak in a matter

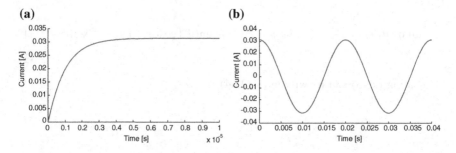

Fig. 2.7 Current in the load when energising at zero voltage. **a** First 10 μs. **b** First 40 ms

of micro-seconds (remember that the variation of a sinus function magnitude around the maximum is very slow).

Figure 2.7 shows the energisation of a RC load with a resistance of 1 Ω and a capacitance of 1 μF connected to a 100 V-peak voltage source. Figure 2.7a zooms the first 10 μs of the energisation and shows that the current reaches the peak value in just 5 μs and that after this instant, only the steady-state current is present.

Energisation at peak voltage:
The energisation of the RC load for zero voltage is equivalent to having $\theta = \pm 90°$. Thus, (2.45) can be simplified into (2.47).

$$I(t) = \pm \frac{V_P}{\sqrt{R^2 + \left(\frac{1}{\omega C}\right)^2}} \left(\cos\left(\omega t + \frac{\pi}{2}\right) + \frac{1}{\omega RC} e^{-\frac{1}{RC}t} \right) \tag{2.47}$$

The analysis indicated that the current jumps from 0 A to current approximately equal to V_P/R instantaneously. Some readers may now be a little confused as they were taught that an instantaneous current change is impossible.

At this point, it is a good idea to distinguish between the electrical and the magnetic fields. A capacitor is an element that stores energy in the electric field, whereas an inductor is an element that stores energy in the magnetic field.

A change in the electric field requires a change in the voltage or charge which is opposed by a current. Thus, an instantaneous change of the voltage would require infinite current, something which is impossible because it would require infinite power. In other words, there must be conservation of charge and the voltage may be continuous at the capacitor.

As an example, think about the voltage-current relation of a capacitor (2.48). If the voltage changes suddenly, like in the connection of the capacitor to an ideal voltage source, the value of dV/dt would be infinite and so also would be the current.

$$I(t) = C \frac{dV}{dt} \tag{2.48}$$

Fig. 2.8 Current (*solid line*) and voltage in the capacitor (*dashed line*) during the first 10 μs of the energisation of an RC load

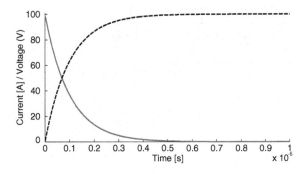

A similar situation happens in a magnetic field, where a change in the current is opposed by an electromotive force (emf), i.e., a voltage. Thus, an instantaneous change of the current would require infinite voltage.

Going back to our RC example, we see that there is an instantaneous change of the current, but not to infinite. This difference is a consequence of the law of conservation of charge preciously stated and the presence of the resistance.

The capacitor is initially uncharged, but it starts to charge when the switch is closed. The voltage at the capacitor has to be continuous, meaning that at $t(0^+)$ all the voltage is dropped at the resistor. As a result, the current at $t(0^+)$ is equal to V_P/R. As the capacitor charges, the voltage at the resistor drops, as does the current, until the system reaches steady-state conditions.

Figure 2.8 shows the current and voltage in the capacitor for an energisation at peak voltage of the RC load used in the previous example. Notice how the voltage at the capacitor raises from zero to approximately 100 V, while the current decreases to the steady-state peak value.

In this particular case, the resistance is 1 Ω and the voltage at the resistor is equal to the current. Therefore, the summation of the two curves at any given instant is equal to the voltage in the source for this particular case.

A capacitor bank strongly resembles a RC circuit and a large current will be present in the energisation of a capacitor bank, if no extra precautions are taken. This current is called inrush current and it has both a high magnitude and a high frequency.

However, a real circuit also has some inductance, which reduces the current magnitude and frequency and more importantly assures the continuity of the current, i.e., there is no longer the current jump that is present in a RC circuit.

2.3.2 The Importance of the Time-Step

We saw that the energisation transient of an RC load is damped in just some micro-seconds. Thus, special care regarding the time-step is necessary when simulating this phenomenon in an EMTP-type software.

Fig. 2.9 Current in the in the capacitor during the first 10 µs for different time steps. *Solid line* 0.01 µs, *dashed line* 0.1 µs, *dotted line* 1 µs

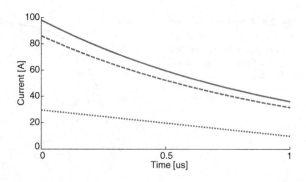

Figure 2.9 shows the current in the RC load during the energisation for different simulation time-steps in EMTDC/PSCAD. The peak current is different for all the three cases and the smaller the time-step, the higher the peak current.

The time constant of this particular example is 1 µs. Thus, the peak current of the 1 µs is approximately 30 A, remembering that the time constant is the time necessary for the current to decay 63.2%.

This is a simple example of the importance of choosing a good time-step and how sometimes it is necessary to use very small time-steps. However, we want to have the time-step as large as possible in order to reduce the simulation running time. Thus, it is necessary to learn how to choose the right time-step in function of the expected phenomenon.

2.3.3 Summary

In this section, we analysed the energisation of an RC load. We saw another example of the use of the Laplace transform, how the transient changes in function of the switching instant and how the transient current and the frequency may be very high when energising the load at peak voltage.

Contrary to the RL load, the transient is "smoother" when energising at zero voltage and more problematic when energising at peak voltage.

Finally, we saw how the time-step influences the results and how it is important to choose the appropriate time-step.

2.4 Switching of RLC Circuits

The final example of circuits with lumped-parameters is the RLC circuit. The RLC circuit is a second-order circuit whose behaviour is more complex than in the previous two examples, as both electric and magnetic fields are present.

The conjugation of these three elements can describe, up to some degree, many of the electrical equipment that exists in real life. As an example, a capacitor bank can be described as just a resistor and a capacitor, but it would never be connected to an ideal voltage source. In reality there is always some inductance between the two elements, making the circuit similar to an RLC circuit. Another example is the pi-model used to represent overhead lines or cables, which we will use in future chapters.

2.4.1 DC Source

A series RLC circuit connected to a DC voltage source is described by (2.49), which can be written in the frequency domain (2.50). Note: For a parallel RLC circuit, see the exercises.

$$V = RI + L\frac{dI}{dt} + \frac{1}{C}\int I dt \tag{2.49}$$

$$\frac{d(V_P)}{dt} = R\frac{dI}{dt} + L\frac{d^2I}{dt^2} + \frac{I}{C} \Leftrightarrow 0 = \frac{d^2I}{dt^2} + \frac{R}{L}\frac{dI}{dt} + \frac{I}{LC} \tag{2.50}$$

A series RLC circuit connected to a DC source has only the transient component of the current, whereas the steady-state current is zero; remember that the capacitor is like an open circuit when in the presence of a DC current. The transient component is a result of an exchange of energy between the capacitor and the inductor that it is eventually damped by the resistance.

Equation (2.50) is a homogeneous differential equation and it can easily be solved without using the Laplace Transform. The derivatives are replaced by λ and (2.50) is replaced by (2.51). The roots of (2.51) are calculated (2.52) and replaced in the general solution of a homogeneous differential equation (2.53).

$$0 = \lambda^2 + \frac{R}{L}\lambda + \frac{1}{LC} \tag{2.51}$$

$$\lambda_{1,2} = -\frac{R}{2L} \pm \sqrt{\left(\frac{R}{2L}\right)^2 - \frac{1}{LC}} \tag{2.52}$$

$$I(t) = C_1 e^{\lambda_1 t} + C_2 e^{\lambda_2 t} \tag{2.53}$$

The roots solution ($\lambda_{1,2}$) can be one of three types, each with a different type of solution.

Two distinct real roots: $(R/2L)^2 > 1/(LC)$ –> Overdamped circuit

Two complex conjugate roots: $(R/2L)^2 < 1/(LC)$ –> Underdamped circuit (Oscillating)

A double root: $(R/2L)^2 = 1/(LC)$ –> Critically damped circuit.

It is still necessary to calculate the values of C_1 and C_2, which depend on the system initial conditions. There are two variables, meaning that two equations are needed (2.54).

$$\begin{cases} I(0) = C_1 + C_2 \\ \dot{I}(0) = \lambda_1 C_1 + \lambda_2 C_2 \end{cases} \tag{2.54}$$

The circuit is de-energised prior to the switching and the value of $I(0)$ is simply zero.

For calculating the value of $\dot{I}(0)$, we must do some deductions. An RLC circuit has both capacitance and inductance and it is not possible to have a sudden change of the voltage because of the capacitor, or of the current because of the inductor. As the current cannot change instantaneously, its value is 0 A at $t = 0^+$ and there is no voltage drop at the resistor. Therefore, all the voltage is dropped in the inductor (2.55).

$$V(0^+) = L\frac{dI}{dt}\bigg|_{t=0^+} \Rightarrow \dot{I}(0) = \frac{V}{L} \tag{2.55}$$

Substituting in (2.54), (2.56) is obtained.

$$\begin{cases} 0 = C_1 + C_2 \\ \frac{V}{L} = \lambda_1 C_1 + \lambda_2 C_2 \end{cases} \tag{2.56}$$

Example:
In the following example, the following three possible scenarios are considered:

1. Overdamped circuit: $V = 100$ V; $L = 0.1$ H; $C = 1$ µF; $R = 1,000$ Ω
2. Underdamped circuit: $V = 100$ V; $L = 0.1$ H; $C = 1$ µF; $R = 100$ Ω
3. Critically damped circuit; $V = 100$ V; $L = 0.1$ H; $C = 1$ µF; $R = 633$ Ω.

The first step is to calculate the roots' values for each case:

1. $\lambda_1 = -1273$; $\lambda_2 = -8873$;
2. $\lambda_{1,2} = -500 \pm j3123$
3. $\lambda_{1,2} \sim 3161$ (in reality, there is a small difference between the roots, as R is an irrational number).

The next step is to calculate the value of the constants C_1 and C_2. The initial current is zero, therefore, $C_1 = -C_2$ for all the cases.

1, $C_1 = -C_2 = 0.1291$
2. $C_1 = -C_2 = -j0.1601$
3. $C_1 = -C_2 = 341$.

Fig. 2.10 Current during the energisation of an RLC circuit. *Solid line* overdamped circuit, *dashed line* underdamped circuit, *dotted line* critically damped circuit

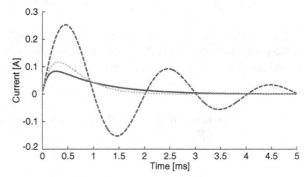

Figure 2.10 shows the transient current for all the three scenarios:

1. The current increases up to a peak value and it damps to zero;
2. The current oscillates around zero, decaying as the time advances. In an LC circuit, the current continues to oscillate at approximately the same frequency, but it is not damped;
3. Identical to scenario 1, but it is the limit case. If the resistance was smaller, and the inductance and capacitance were the same, there would be an oscillation.

2.4.2 AC Source

One would think that the transient obtained when connecting an RLC load to an AC source is completely different from connecting the load to an equivalent DC source (i.e., the same peak amplitude). However, we will see that this is not necessarily true.

We start by writing the equation describing the system on the time domain (2.57) and simplify it to (2.58) and apply the Laplace Transform (2.59).

$$V_P \sin(\omega t + \theta) = RI + L\frac{dI}{dt} + \frac{1}{C}\int I dt \tag{2.57}$$

$$\frac{d(V_P\sin(\omega t + \theta))}{dt} = R\frac{dI}{dt} + L\frac{d^2I}{dt^2} + \frac{I}{C} \Leftrightarrow V_P\omega\cos(\omega t + \theta) = L\frac{d^2I}{dt^2} + R\frac{dI}{dt} + \frac{I}{C} \tag{2.58}$$

$$V_P\omega\left(\frac{s\cos(\theta)}{s^2 + \omega^2} - \frac{\omega\sin(\theta)}{s^2 + \omega^2}\right) = I\left(s^2L + sR + \frac{1}{C}\right) - sLI(0) - L\dot{I}(0) - RI(0) \tag{2.59}$$

We already know from the previous sections that the current at t_0 is zero because of the continuity of the current, and that the derivative of the current at the same instant is given by $V_P(0)/L$. Thus, (2.59) becomes (2.60).

$$V_P\omega\left(\frac{s\cos(\theta)}{s^2 + \omega^2} - \frac{\omega\sin(\theta)}{s^2 + \omega^2}\right) = I\left(s^2L + sR + \frac{1}{C}\right) - L\frac{V_P(0)}{L} \tag{2.60}$$

At this point, we may solve the system using the Laplace transform as described in the previous sections, but this is unnecessary. The current is a summation of the forced and homogeneous components (2.61).

$$I(t) = I_f(t) + I_h(t) \tag{2.61}$$

The forced component is the steady-state component and it is easy to obtain by using phasors (2.62).

$$I_f(t) = \frac{V_P}{\sqrt{R^2 + \left(\omega L - \frac{1}{\omega C}\right)^2}} \sin\left(\omega t + \theta - \tan^{-1}\left(\frac{\omega L - \frac{1}{\omega C}}{R}\right)\right) \tag{2.62}$$

The homogeneous component is similar to the one obtained when using a DC source, with the difference that the value of $V_P(0)$ depends on the switching instant and that the derivative of the voltage is no longer zero. The relation (2.63) continues to be valid, but the values of $I_h(0)$ and $\dot{I}_h(0)$ need to be calculated. The premises used for the DC source example continue to be valid; the initial current has to be zero, thus (2.64), and the value of the derivative of the current at $t = 0$ is still given by having all the voltage dropping in the inductor. Thus, the derivative of the homogeneous current component is given by (2.65). The variables λ_1 and λ_2 continue to be calculated by (2.52).

$$\begin{cases} I_h(0) = C_1 + C_2 \\ \dot{I}_h(0) = \lambda_1 C_1 + \lambda_2 C_2 \end{cases} \tag{2.63}$$

$$I(0) = 0 \Leftrightarrow I_h(0) = -I_f(0) \tag{2.64}$$

$$\dot{I}(0) = \dot{I}_h(0) + \dot{I}_f(0) \Rightarrow \dot{I}_h(0)$$

$$= \left(\frac{V_P \omega}{\sqrt{R^2 + \left(\omega L - \frac{1}{\omega C}\right)^2}} \cos\left(\theta - \tan^{-1}\left(\frac{\omega L - \frac{1}{\omega C}}{R}\right)\right)\right) - \left(\frac{V_P}{L}\sin(\theta)\right)$$

$$\tag{2.65}$$

Replacing (2.64) and (2.65) in (2.63), (2.66) is obtained for C_1 and C_2. Replacing the results in (2.61), the general expression of the current in the RLC load when connected to an AC source is obtained (2.67).

$$\begin{cases} C_1 = \frac{\dot{I}_h(0) + I_f(0)\lambda_2}{\lambda_1 - \lambda_2} \\ C_2 = \frac{\dot{I}_h(0) + I_f(0)\lambda_1}{\lambda_2 - \lambda_1} \end{cases} \tag{2.66}$$

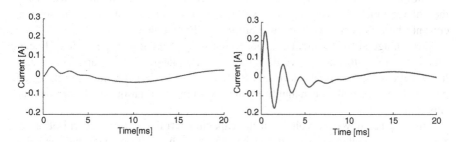

Fig. 2.11 Current during the energisation of the RLC load. *Left* energisation at $\theta = 0°$, *right* energisation at $\theta = 90°$

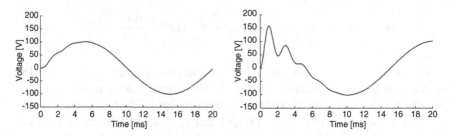

Fig. 2.12 Voltage in the capacitor (*VC*) during the energisation of the RLC load. *Left* energisation at $\theta = 0°$, *right* energisation at $\theta = 90°$

$$I(t) = \frac{V_P}{\sqrt{R^2 + \left(\omega L - \frac{1}{\omega C}\right)^2}} \sin\left(\omega t + \theta - \tan^{-1}\left(\frac{\omega L - \frac{1}{\omega C}}{R}\right)\right)$$
$$+ \frac{\dot{I}_h(0) + I_f(0)\lambda_2}{\lambda_1 - \lambda_2} e^{\lambda_1 t} + \frac{\dot{I}_h(0) + I_f(0)\lambda_1}{\lambda_2 - \lambda_1} e^{\lambda_2 t} \tag{2.67}$$

Figure 2.11 shows the current during an energisation at zero voltage and peak voltage respectively. The simulation parameters are $V_P = 100$ V; $L = 0.1$ H; $C = 1$ μF; $R = 100$ Ω.

Notice that the waveform for the energisation at peak voltage is very similar, regarding both shape and magnitude, to the one obtained when energising the same load using a DC source of equal magnitude. The homogeneous component depends on the voltage value at the energisation instant and the load initial conditions.

The voltage initial conditions and the load initial conditions are the same in both cases. The small difference between the two cases is present because the forced component is not zero at $t = 0$. However, as this component is close to zero, the difference between the two waveforms is very small.

It should also be noticed that the homogeneous waveform is influenced by the load parameters in the same way as for the energisation with a DC source. As an example, an overdamped oscillation would be present if the resistor has of 1,000 Ω.

Until now, we have been focusing on the current behaviour and not considering the voltage much. Yet, in the same way that there is a transient waveform for the current, there is also a transient waveform for the voltage.

The voltage at the terminals of each element is a function of the current: linear function if a resistor, derivative if an inductor and integrative if a capacitor. Thus, the voltage at the element's terminal is expected to be larger when energising at peak voltage, as both the magnitude and the current variation are larger in this situation.

Figure 2.12 shows the voltage at the capacitor terminals for an energisation at zero and peak voltage, where it can be seen that the voltage magnitude is substantially larger for the second case.

We will return to this topic in the next chapters and understand better how the switching instant may be a determining parameter when switching a cable.

2.4.3 Summary

In this section, we analysed an RLC circuit and saw how its behaviour is strongly influenced by both the switching instant and the load parameters. We saw how the current or voltage can be divided into two types, the forced and homogeneous regime. The forced regime consists of the system in steady-state condition, whereas the homogenous regime is a transient condition, the total current/voltage is the summation of both regimes.

The equations were solved without the use of the Laplace transform in order to show the reader other possibilities. However, the system can still be solved using the Laplace transform. As a matter of fact, this is one of the exercises proposed next, and the solution may be found online.

2.5 Exercises

1. Obtain the current expression for the RLC circuit of Sect. 2.4 (DC source) using the Laplace Transform.

$$A: I = \frac{V}{L} \left(\frac{2\sqrt{\left(\frac{R}{2L}\right)^2 - \left(\frac{1}{\sqrt{LC}}\right)^2}}{s - \left(-\frac{R}{2L} - \sqrt{\left(\frac{R}{2L}\right)^2 - \left(\frac{1}{\sqrt{LC}}\right)^2}\right)} - \frac{2\sqrt{\left(\frac{R}{2L}\right)^2 - \left(\frac{1}{\sqrt{LC}}\right)^2}}{s - \left(-\frac{R}{2L} + \sqrt{\left(\frac{R}{2L}\right)^2 - \left(\frac{1}{\sqrt{LC}}\right)^2}\right)} \right)$$

2. Repeat exercise 1, but for a parallel RLC circuit, for both a DC and AC voltage sources.

$$A: I(t) = \frac{V}{R} + \frac{V}{L}t + VC\delta(t); \quad I = \frac{V\sin(\omega t)}{R} + \frac{V}{\omega L} - \frac{V}{\omega L}\cos(\omega t) + \omega CV\cos(\omega t)$$

3. Obtain the expression for the transient of the voltage at the receiving end of a line connected to an AC source using a pi-model. The energisation is made at peak voltage and the values are the following: $R = 0.62\ \Omega$; $L = 44.7$ mH; $C = 3.9\ \mu F$; $V_P = 100$ kV.

$$A: V_2 = \frac{1}{1.95\times 10^{-6}}\left(0.2\cos(\omega t) + e^{-6.94t}(4\times 10^{-4}\sin(3387t) + 0.2\cos(3387t))\right)$$

4. For the same pi-model obtain the expression of the current at the sending end of the line.

$$A: I(t) = -61.26\sin(\omega t) - 61.75\sin(\omega t) - e^{-6.94t}(666\sin(3387t))$$

References and Further Reading

1. Kreyszig Erwin (1988) Advanced engineering mathematics, 6th edn. Wiley, New York
2. Greenwood Allan (1991) Electrical transients in power systems, 2nd edn. Wiley, New York
3. David Irwin J, Mark Nelms R (2008) Basic engineering circuit analysis, 9th edn. Wiley, New York

Chapter 3
Travelling Waves

3.1 Introduction

Before we start analysing the different transient phenomena associated to the use of the HVAC cables, we should understand the mathematics and physics behind the models used in the simulations.

This chapter recaps some of the travelling wave classic theory and delves deep into the modelling of underground cables, by explaining how to calculate the series impedance and admittance matrices. The chapter also introduces the modal equations that are essential for the understating of some electromagnetic transient phenomena.

3.2 The Telegraph Equations

The modelling of a line, either a cable or an overhead line, can be obtained by applying the Kirchhoff's laws to the circuit shown in Fig. 3.1, which represents an infinitesimal section of the line. By doing this, Eqs. (3.1) and (3.2) are obtained. Notice that it is necessary to use partial derivatives as the current and the voltage are functions of both time and distance.

$$V(x, t) - V(x + dx, dt) = Rdx \cdot I(x, t) + Ldx \frac{\partial I(x, t)}{\partial t} \tag{3.1}$$

$$I(x, t) - I(x + dx, t) = Gdx \cdot V(x, t) + Cdx \frac{\partial V(x, t)}{\partial t} \tag{3.2}$$

F. F. da Silva and C. L. Bak, *Electromagnetic Transients in Power Cables*,
Power Systems, DOI: 10.1007/978-1-4471-5236-1_3,
© Springer-Verlag London 2013

Fig. 3.1 Equivalent single-phase circuit of a length dx cable

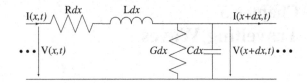

3.2.1 Time Domain

For simplicity, we start by considering the line as being lossless. Thus, Eqs. (3.1) and (3.2) are written as (3.3) and (3.4), also called telegrapher's equations for lossless lines.

$$V(x, t) - V(x + dx, dt) = Ldx\frac{\partial I(x, t)}{\partial t} \Leftrightarrow \frac{\partial V(x, t)}{\partial x} = -Ldx\frac{\partial I(x, t)}{\partial t} \qquad (3.3)$$

$$I(x, t) - I(x + dx, t) = Cdx\frac{\partial V(x, t)}{\partial t} \Leftrightarrow \frac{\partial I(x, t)}{\partial x} = -Cdx\frac{\partial V(x, t)}{\partial t} \qquad (3.4)$$

Equations can be further developed to (3.5) and (3.6), whose typical solution is given in Eqs.(3.7) and (3.8).

$$\frac{\partial^2 V(x, t)}{\partial x^2} = LC\frac{\partial^2 V(x, t)}{\partial t^2} \qquad (3.5)$$

$$\frac{\partial^2 I(x, t)}{\partial x^2} = LC\frac{\partial^2 I(x, t)}{\partial t^2} \qquad (3.6)$$

$$V(x, t) = V^+\left(t - \sqrt{LC}x\right) + V^-\left(t + \sqrt{LC}x\right) \qquad (3.7)$$

$$I(x, t) = I^+\left(t - \sqrt{LC}x\right) + I^-\left(t + \sqrt{LC}x\right) \qquad (3.8)$$

At this point, it is important to do a physical analysis of the equations in order to understand what they are describing. We can see that both the voltage and the current have two components V^+/I^+ and V^-/I^-, which are respectively the *forward* and *backward* waves. As the names indicate, the forward wave represents the wave propagating in the positive direction of the line (typically from the sending end to the receiving end) and the backward wave represents the wave propagating in the negative direction. The total wave is the summation of these two components at a given instant and point of the line as given by the superposition of both components.

The physical meaning of these two components is rather simple. Imagine a generator sending a wave into a line that is open. The wave propagates itself (as forward wave) in the line until the end where it is reflected back; after this instant, the wave in the line is the summation of the wave that is being sent by the generator, the forward wave and its reflection, the backward wave.

By now, some readers will have noticed that the propagation speed of a lossless wave is given by $1/\sqrt{LC}$. In other words, the wave speed depends only on the characteristics of the line and it is independent on the length and sending signal. Thus, if two lines with different parameters per length are connected to each other, the wave propagates at different speeds in each of the lines. This is very important when simulating transient in a hybrid line consisting of a cable, where the wave coaxial speed is approximately 180 m/μs, and an OHL where the wave speed is approximately 280 m/μs.

In the previous sentence were stated typical wave speeds for a cable and an OHL and one may wonder if these values are always valid. As a matter of fact, they are valid for most of the cases. In the previous chapter, it was explained how to calculate the inductance and capacitance. The inductance and capacitance in an isolated conductor are calculated using respectively Eqs. (3.9) and (3.10), where a and b are respectively the radius of the conductor and of the insulation.

$$C = \frac{2\pi\varepsilon}{\ln\left(\frac{b}{a}\right)} \tag{3.9}$$

$$L = \frac{\mu}{2\pi}\ln\left(\frac{b}{a}\right) \tag{3.10}$$

Substituting these values in the speed Eq. (3.11) is obtained.

$$v = \frac{1}{\sqrt{LC}} \Leftrightarrow v = \frac{1}{\sqrt{\mu\varepsilon}} \Leftrightarrow v \simeq \frac{1}{\sqrt{\mu_0 2.5\varepsilon_0}} \Leftrightarrow v \simeq \frac{1}{\sqrt{2.5}}c \Leftrightarrow v \simeq 190\,\mathrm{m/\mu s} \tag{3.11}$$

3.2.2 Frequency Domain

We have seen how the waves behave in the time domain and we can now do the same for the frequency domain. Equations (3.1) and (3.2) can be converted into the frequency domain as done in Eqs. (3.12) and (3.13). Notice that the equations are now for steady-state conditions and that the line is no longer lossless.

$$-\frac{dV(x,\omega)}{dx} = (R(\omega) + j\omega L(\omega)) \cdot I(x,\omega) \tag{3.12}$$

$$-\frac{dI(x,\omega)}{dx} = (G(\omega) + j\omega C(\omega)) \cdot V(x,\omega) \tag{3.13}$$

The derivation of Eqs. (3.12) and (3.13) leads to Eqs. (3.14) and (3.15), respectively.

$$\frac{d^2V(x, \omega)}{dx^2} = (R(\omega) + j\omega L(\omega)) \cdot \left(-\frac{dI(x, \omega)}{dx} \right)$$

$$\Leftrightarrow \frac{d^2V(x, \omega)}{dx^2} = ((R(\omega) + j\omega L(\omega))(G(\omega) + j\omega C(\omega)))V(x, \omega) \tag{3.14}$$

$$\frac{d^2I(x, \omega)}{dx^2} = (G(\omega) + j\omega C(\omega)) \cdot \frac{dV(x, \omega)}{dx}$$

$$\Leftrightarrow \frac{d^2I(x, \omega)}{dx^2} = ((G(\omega) + j\omega C(\omega))(R(\omega) + j\omega L(\omega)))I(x, \omega) \tag{3.15}$$

The characteristic impedance and propagation constant of a general line are defined by Eqs. (3.16) and (3.17), respectively. These two entities are important, because they allow both to simplify the equation and to provide important information on the behaviour of the waves in a line.

$$Z_0(\omega) = \sqrt{\frac{R(\omega) + j\omega L(\omega)}{G(\omega) + j\omega C(\omega)}} \tag{3.16}$$

$$\gamma(\omega) = \sqrt{(R(\omega) + j\omega L(\omega)) \cdot (G(\omega) + j\omega C(\omega))} \tag{3.17}$$

The propagation constant (3.17) can be used in Eqs. (3.14) and (3.15), which are simplified to (3.18) and (3.19), respectively.

$$\frac{d^2V(x, \omega)}{dx^2} = \gamma^2(\omega)V(x, \omega) \tag{3.18}$$

$$\frac{d^2I(x, \omega)}{dx^2} = \gamma^2(\omega)I(x, \omega) \tag{3.19}$$

Equations (3.18) and (3.19) are ordinary differential equations whose typical solution are given by respectively (3.20) and (3.21), where A and B are constants that have to be calculated using the system initial conditions, typically $x = 0$ for the sake of simplicity.

$$V(x, \omega) = A_1 e^{-\gamma x} + A_2 e^{\gamma x} \tag{3.20}$$

$$I(x, \omega) = B_1 e^{-\gamma x} + B_2 e^{\gamma x} \tag{3.21}$$

Before calculating the constant, it is important to refer an important relation between the voltage and the current. We know that a relation between voltage and current is given by an impedance, which in the case of a line is the characteristic impedance given in (3.16). Thus, the current Eq. (3.21) can be written as (3.22) and the same constants are used for both voltage and current.

$$I(x, \omega) = \frac{A_1}{Z_0} e^{-\gamma x} - \frac{A_2}{Z_0} e^{\gamma x} \tag{3.22}$$

The constant can now be easily calculated by using the voltage and current in the sending end (3.23).

$$\begin{cases} V_S = A_1 + A_2 \\ I_S = \frac{A_1 - A_2}{Z_0} \end{cases} \Leftrightarrow \begin{cases} A_1 = \frac{V_S + Z_0 I_S}{2} \\ A_2 = \frac{V_S - Z_0 I_S}{2} \end{cases} \tag{3.23}$$

The equations can be further developed using hyperbolic functions. As a result, the voltage and the current can be written as respectively (3.24) and (3.25).

$$V(x, \omega) = V_S \cosh(x\gamma(\omega)) - Z_0 I_S \sinh(x\gamma(\omega)) \tag{3.24}$$

$$I(x, \omega) = -\frac{V_S}{Z_0} \sinh(x\gamma(\omega)) + I_S \cosh(x\gamma(\omega)) \tag{3.25}$$

In the receiving end of the line x=l, where l is the line length, and (3.24) and (3.25) become respectively (3.26) and (3.27). This is also called the line exact equation; however, one should keep in mind that the results are only accurate for the chosen frequency.

$$V_R(\omega) = V_S \cosh(l\gamma(\omega)) - Z_0 I_S \sinh(l\gamma(\omega)) \tag{3.26}$$

$$I_R(\omega) = -\frac{V_S}{Z_0} \sinh(l\gamma(\omega)) + I_S \cosh(l\gamma(\omega)) \tag{3.27}$$

3.3 Impedance and Admittance Matrices of a Cable

We have seen how to use the telegraph equations for a single-phase conductor. The equations can with some changes still be applied for three-phase cable in steady-state conditions, but special care is required.

However, a cable may have several conductors per phase, which requires some changes in the procedures. The first big change is in the calculation of the impedance and admittance of the cable, which is explained in the next pages.

3.3.1 Both Ends Bonded Cable

HVAC cables are commonly installed using three single-core cables. In this situation, the cable series impedance becomes a 6×6 matrix or a 9×9 matrix depending whether the cable has armour or not.

The calculations are similar for both cases and we start by considering a 9×9 matrix as shown in Eq. (3.28). The matrix contains both the self-impedance and the mutual impedances. The first column and the first row of the matrix represent the core/conductor of Phase A, the second column and row represent the screen of

Phase A, the third column and row represent the armour of Phase A, whereas the others columns and rows represent Phases B and C in the same order. Thus, the main diagonal is the self-impedance of the cores, screens and armours, whereas the upper and lower triangles are the mutual impedances, resulting in a symmetrical matrix.

$$
[Z] = \begin{bmatrix}
Z_{C1C1} & Z_{C1S1} & Z_{C1A1} & Z_{C1C2} & Z_{C1S2} & Z_{C1A2} & Z_{C1C3} & Z_{C1S3} & Z_{C1A3} \\
Z_{S1C1} & Z_{S1S1} & Z_{S1A1} & Z_{S1C2} & Z_{S1S2} & Z_{S1A2} & Z_{S1C3} & Z_{S1S3} & Z_{S1A3} \\
Z_{A1C1} & Z_{A1S1} & Z_{A1A1} & Z_{A1C2} & Z_{A1S2} & Z_{A1A2} & Z_{A1C3} & Z_{A1S3} & Z_{A1A3} \\
Z_{C2C1} & Z_{C2S1} & Z_{C2A1} & Z_{C2C2} & Z_{C2S2} & Z_{C2A2} & Z_{C2C3} & Z_{C2S3} & Z_{C2A3} \\
Z_{S2C1} & Z_{S2S1} & Z_{S2A1} & Z_{S2C2} & Z_{S2S2} & Z_{S2A2} & Z_{S2C3} & Z_{S2S3} & Z_{S2A3} \\
Z_{A2C1} & Z_{A2S1} & Z_{A2A1} & Z_{A2C2} & Z_{A2S2} & Z_{A2A2} & Z_{A2C3} & Z_{A2S3} & Z_{A2A3} \\
Z_{C3C1} & Z_{C3S1} & Z_{C3A1} & Z_{C3C2} & Z_{C3S2} & Z_{C3A2} & Z_{C3C3} & Z_{C3S3} & Z_{C3A3} \\
Z_{S3C1} & Z_{S3S1} & Z_{S3A1} & Z_{S3C2} & Z_{S3S2} & Z_{S3A2} & Z_{S3C3} & Z_{S3S3} & Z_{S3A3} \\
Z_{A3C1} & Z_{A3S1} & Z_{A3A1} & Z_{A3C2} & Z_{A3S2} & Z_{A3A2} & Z_{A3C3} & Z_{A3S3} & Z_{A3A3}
\end{bmatrix}
$$

Core 1 Screen 1 Arm. 1 Core 2 Screen 2 Arm. 2 Core 3 Screen 3 Arm. 3

$$(3.28)$$

The matrix can be further simplified by doing some assumptions and approximations:

1. The three cables are equal. Thus, $Z_{C1C1} = Z_{C2C2} = Z_{C3C3}$, $Z_{S1S1} = Z_{S2S2} = Z_{S3S3}$, $Z_{A1A1} = Z_{A2A2} = Z_{A3A3}$, $Z_{C1S1} = Z_{C2S2} = Z_{C3S3}$ and $Z_{C1A1} = Z_{C2A2} = Z_{C3A3}$

2. The mutual inductance between phases is a function of the distance. The distance between the core of Phase A and the core of Phase B is practically equal to the distance between the core of Phase A and the screen of Phase B; the same is true for the other phases. Therefore, $Z_{C1C2} \approx Z_{C1S2} \approx Z_{S1C2} \approx Z_{S1S2}$, $Z_{C1C3} \approx Z_{C1S3} \approx Z_{S1C3} \approx Z_{S1S3}$, $Z_{C2C3} \approx Z_{C2S3} \approx Z_{S2C3} \approx Z_{S2S3}$, and the same reasoning for the armour.

As a result, the series impedance matrix (3.28) can be simplified to (3.29), where it is noticeable that the entries of the matrix can be divided into:

- $Z_{11/22/33}$: the self-impedance of the core/screen/armour;
- $Z_{12/13/23}$: the mutual impedance between core-screen/core-armour/screen-armour;
- $Z_{gm12/13/23}$: the ground mutual impedance between phases.

$$[Z] = \begin{bmatrix} Z_{11} & Z_{12} & Z_{13} & Z_{gm12} & Z_{gm12} & Z_{gm12} & Z_{gm13} & Z_{gm13} & Z_{gm13} \\ Z_{12} & Z_{22} & Z_{23} & Z_{gm12} & Z_{gm12} & Z_{gm12} & Z_{gm13} & Z_{gm13} & Z_{gm13} \\ Z_{13} & Z_{23} & Z_{33} & Z_{gm12} & Z_{gm12} & Z_{gm12} & Z_{gm13} & Z_{gm13} & Z_{gm13} \\ Z_{gm12} & Z_{gm12} & Z_{gm12} & Z_{11} & Z_{12} & Z_{13} & Z_{gm23} & Z_{gm23} & Z_{gm23} \\ Z_{gm12} & Z_{gm12} & Z_{gm12} & Z_{12} & Z_{22} & Z_{23} & Z_{gm23} & Z_{gm23} & Z_{gm23} \\ Z_{gm12} & Z_{gm12} & Z_{gm12} & Z_{13} & Z_{23} & Z_{33} & Z_{gm23} & Z_{gm23} & Z_{gm23} \\ Z_{gm13} & Z_{gm13} & Z_{gm13} & Z_{gm23} & Z_{gm23} & Z_{gm23} & Z_{11} & Z_{12} & Z_{13} \\ Z_{gm13} & Z_{gm13} & Z_{gm13} & Z_{gm23} & Z_{gm23} & Z_{gm23} & Z_{12} & Z_{22} & Z_{23} \\ Z_{gm13} & Z_{gm13} & Z_{gm13} & Z_{gm23} & Z_{gm23} & Z_{gm23} & Z_{13} & Z_{23} & Z_{33} \end{bmatrix}$$

$$(3.29)$$

A current flowing in the core or in the screen needs to form a closed loop, or else there is no current circulating, similar to what happens in any circuit.

The cable we are describing has for each phase three conductors (core, screen and armour) and three current loops. In the first loop, the current flows in the core and returns in the screen, in the second loop, the current flows in the screen and returns in the armour whereas in the third loop, the current flows in the armour and returns in the ground.

Each of these loops is the summation of several impedances, which are shown and explained next having as base the cable section shown in Fig. 3.2

Conductor Series Impedance

As the name indicates, this parameter is the internal impedance of the core (3.30). It depends on the resistivity of the conductor, the radius of the conductor and the penetration depth. Notice that the formula is frequency dependent and takes in account the skin effect.

The equation assumes a solid conductor, but very often a cable has a stranded conductor. Thus, the resistivity has to be corrected using Eq. (3.31), where A_C is the cross-sectional area as given in the datasheet.

Fig. 3.2 Cross-section of a single-core cable

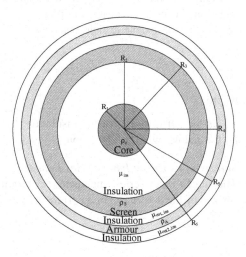

$$Z_{\text{Couter}}(\omega) = \frac{\rho'_C m_C}{2\pi R_1} \cdot \frac{J_0(m_C R_1)}{J_1(m_C R_1)} \tag{3.30}$$

where,

ρ_c is the resistivity of the conductor,
m_c is the reciprocal of the complex penetration depth for the conductor, and given by (3.32),
R_1 is the radius over the conductor,
$J_n(x)$ is the Bessel function of x, of first kind and order n.

$$\rho'_C = \rho_C \frac{\pi R_1}{A_C} \tag{3.31}$$

$$m_C = \sqrt{\frac{j\omega\mu}{\rho'_C}} \tag{3.32}$$

The use of the Bessel functions provides the most accurate results, but at the cost of a larger complexity. An approximation of (3.30) is given in (3.33), where k is an arbitrary constant used to optimise the formula for low frequencies, typically equal to 0.777.

$$Z_{\text{Couter}}(\omega) = \frac{\rho'_C m_C}{2\pi R_1} \cdot \coth(m_C R_1 k) + \frac{\rho'_C}{\pi R_1^2}\left(1 - \frac{1}{2k}\right) \tag{3.33}$$

Insulation Series Impedance
The insulation series impedance is a result of the time-varying magnetic field in the inner insulation of the cable, and it is calculated through (3.34). The name of this term is misleading as there is no current flowing in the insulation and consequently no impedance directly associated the insulation, as can be seen in the equation. In reality, this term is function of the currents flowing in the core and screen and the magnetic field between them.

$$Z_{\text{CSinsul}}(\omega) = \frac{j\omega\mu_{\text{ins}}}{2\pi}\ln\left(\frac{R_2}{R_1}\right) \tag{3.34}$$

where,

μ_{ins} is the permeability of the insulation
R_2 the radius over the insulation

Screen and Armour Inner Series Impedance
The screen inner series impedance is associated to the voltage drop on the inner surface of the screen when a current returns via the core. It is calculated by using (3.35).

$$Z_{\text{Sinner}}(\omega) = \frac{\rho_S m_S}{2\pi R_2} \frac{J_0(m_S R_2)K_1(m_S R_3) + K_0(m_S R_2)J_1(m_S R_3)}{J_1(m_S R_3)K_1(m_S R_2) - J_1(m_S R_2)K_1(m_S R_3)} \tag{3.35}$$

where,

ρ_S is the resistivity of the screen,
m_S is the reciprocal of the complex penetration depth for the screen,
R_3 is the radius over the screen,
$K_n(x)$ is the Bessel function of x, of second kind and order n.

Similar to the conductor series impedance, (3.35) can be simplified to (3.36).

$$Z_{\text{Sinner}}(\omega) = \frac{\rho_S m_S}{2\pi R_2} \coth(m_S(R_3 - R_2)) - \frac{\rho_S}{2\pi R_2(R_2 + R_3)} \tag{3.36}$$

The inner series impedance for the armour is obtained using similar expressions (3.37) and (3.38).

$$Z_{\text{Ainner}}(\omega) = \frac{\rho_A m_A}{2\pi R_4} \frac{J_0(m_A R_4)K_1(m_A R_5) + K_0(m_A R_4)J_1(m_A R_5)}{J_1(m_A R_5)K_1(m_A R_4) - J_1(m_A R_4)K_1(m_A R_5)} \tag{3.37}$$

where,

ρ_A is the resistivity of the armour,
m_A is the reciprocal of the complex penetration depth for the armour,
R_4 is the radius over the screen insulation,
R_5 is the radius over the armour.

$$Z_{\text{Ainner}}(\omega) = \frac{\rho_A m_A}{2\pi R_4} \coth(m_A(R_5 - R_4)) - \frac{\rho_A}{2\pi R_4(R_4 + R_5)} \tag{3.38}$$

Screen and Armour Outer Series Impedance
The screen outer series impedance is analogous to the screen inner series impedance, but for the outer surface of the screen instead of the inner surface. In other words, it is associated to the voltage drop on the inner surface of the screen when a current returns through the armour.

The classic formula is given by (3.39) and the approximation by (3.40). Notice that the formulas are very similar to respectively (3.35) and (3.36), and that the only difference is the substitution of the radius R_2 by R_3 in the first term of (3.39).

$$Z_{\text{Souter}}(\omega) = \frac{\rho_S m_S}{2\pi R_3} \frac{J_0(m_S R_3)K_1(m_S R_2) + K_0(m_S R_3)J_1(m_S R_2)}{J_1(m_S R_3)K_1(m_S R_2) - J_1(m_S R_2)K_1(m_S R_3)} \tag{3.39}$$

$$Z_{\text{Souter}}(\omega) = \frac{\rho_S m_S}{2\pi R_3} \coth(m_S(R_3 - R_2)) + \frac{\rho_S}{2\pi R_3(R_2 + R_3)} \tag{3.40}$$

The formula for the armour outer series impedance is similar and it is only necessary to adjust the radiuses, the resistivity and the penetration depth (3.41) and (3.42).

$$Z_{A\text{outer}}(\omega) = \frac{\rho_A m_A}{2\pi R_5} \frac{J_0(m_A R_5) K_1(m_A R_4) + K_0(m_A R_5) J_1(m_A R_4)}{J_1(m_A R_5) K_1(m_A R_4) - J_1(m_A R_4) K_1(m_A R_5)} \tag{3.41}$$

$$Z_{A\text{outer}}(\omega) = \frac{\rho_A m_A}{2\pi R_5} \coth(m_A(R_5 - R_4)) + \frac{\rho_A}{2\pi R_5(R_4 + R_5)} \tag{3.42}$$

Screen and Armour Insulation Series Impedance
Similar to the insulation series impedance, the screen insulation series impedance is a result of the time-varying magnetic field in the inner insulation of the cable and it is calculated through (3.43).

$$Z_{SA\text{insul}}(\omega) = \frac{j\omega \mu_{\text{out\_ins}}}{2\pi} \ln\left(\frac{R_4}{R_3}\right) \tag{3.43}$$

where, $\mu_{\text{out\_ins}}$ is the permeability of the screen insulation.
For the armour the expression is changed to (3.44).

$$Z_{AG\text{insul}}(\omega) = \frac{j\omega \mu_{\text{out2\_ins}}}{2\pi} \ln\left(\frac{R_6}{R_5}\right) \tag{3.44}$$

where,

$\mu_{\text{out2\_ins}}$ is the permeability of the armour insulation,
R_6 the radius over the armour insulation.

Mutual Series Impedance of two loops
The mutual series impedance is associated to two different current loops and equal for both. A first loop where there is a voltage drop in the outer surface of the screen because of a current returning in the core and a second loop where there is a voltage drop in the inner surface of the screen because of a current returning in the armour.

The mutual series impedance is calculated by (3.45), which may be simplified into (3.46).

$$Z_{S\text{mutual}} = \frac{\rho_S}{2\pi R_2 R_3} \frac{1}{J_1(m_S R_3) K_1(m_S R_2) - J_1(m_S R_2) K_1(m_S R_3)} \tag{3.45}$$

$$Z_{S\text{mutual}} = \frac{\rho_S m_S}{\pi(R_2 + R_3)} \operatorname{csch}(m_S(R_3 - R_2)) \tag{3.46}$$

There is also a mutual series impedance associated to the armour. In the first loop, there is a voltage drop in the outer surface of the armour because of a current returning in the screen, whereas in the second loop, there is a voltage drop in the inner surface of the armour because of a current returning in the ground. The equations are similar to the previous ones, except that the radius, the resistivity and the penetration depth are adjusted for the armour (3.47) and (3.48).

$$Z_{\text{Amutual}} = \frac{\rho_A}{2\pi R_4 R_5} \frac{1}{J_1(m_A R_5) K_1(m_A R_4) - J_1(m_A R_4) K_1(m_A R_5)} \quad (3.47)$$

$$Z_{\text{Amutual}} = \frac{\rho_A m_A}{\pi(R_4 + R_5)} \text{csch}(m_A(R_5 - R_4)) \quad (3.48)$$

Earth Series Self Impedance
The earth series self-impedance is the parameter whose calculation arises more difficulties and it is also the one presenting the largest errors. The classic formula based on original developments made by Pollaczek is given by (3.49).

$$Z_{\text{earth}}(\omega) = \frac{\rho_e m_e^2}{2\pi} \left[K_0(m_e R_4) - K_0\left(m_e\sqrt{R_4^2 + 4h^2}\right) + \int_{-\infty}^{+\infty} \frac{e^{-2h\sqrt{m_e^2+\alpha^2}}}{|\alpha| + \sqrt{m_e^2+\alpha^2}} e^{j\alpha R_4} d\alpha \right]$$

$$(3.49)$$

where,

ρ_e is the resistivity of the earth,
m_e is the reciprocal of the complex penetration depth for the earth,
h is the depth of the cable.

The computation of (3.49) is very challenging because of the integral and it is normal to use approximations. Several formulas were proposed through the years; one of the most accurate is the one proposed by Saad, Gaba and Giroux in 1996, which simplifies (3.49)–(3.50).

$$Z_{\text{earth}}(\omega) = \frac{\rho_e m_e^2}{2\pi} \left[K_0(m_e R_4) + \frac{2}{4 + m_e^2 R_4^2} e^{-2hm_e} \right] \quad (3.50)$$

Mutual Earth Impedance
The mutual earth impedance represents the mutual inductance between the cables. The mutual inductance is the result of a voltage drop in the earth adjacent to the cables inducing an e.m.f. in the cables when the current returns through the soil.

The formula used to calculate the impedance is given by (3.51), which can be simplified to (3.52). Notice the resemblance with the formulas used to estimate the earth series self-impedance, something logic as in both cases the current is returning in the soil.

$$Z_{\text{earth\_mutual}}(\omega) = \frac{\rho_e m_e^2}{2\pi} \left[K_0(m_e d) - K_0\left(m_e\sqrt{d^2 + (h_i - h_j)^2}\right) + \int_{-\infty}^{+\infty} \frac{e^{-(h_i+h_j)\sqrt{m_e^2+\alpha^2}}}{|\alpha| + \sqrt{m_e^2+\alpha^2}} e^{j\alpha d} d\alpha \right]$$

$$(3.51)$$

where,

d is the distance between the conductors

$h_{i,j}$ are the buried depths of cables i and j

$$Z_{\text{earth\_mutual}}(\omega) = \frac{\rho_e m_e^2}{2\pi} \left[K_0(m_e d) + \frac{2}{4 + m_e^2 d^2} e^{-2hm_e} \right] \quad (3.52)$$

Configuration of the loops

We now know how to calculate the several impedances in a cable, but not how to use them in the loops.

For simplicity, we start with a single-phase cable and the core-screen loop. The impedances of this loop are equal to the core self-impedance, plus the insulation series impedance and the screen inner impedance (3.53), which are the series of all the impedance in the path travelled by the current. The impedance of the screen-armour loop is similar and it is equal to the screen outer series impedance, plus the screen outer-insulation series impedance and the armour inner impedance (3.54). Finally, the impedance of the armour ground is equal to the armour outer series impedance, plus the armour outer-insulation series impedance and the earth series impedance (3.55). Figure 3.3 shows the equivalent circuit for the three loops.

$$Z_{L\_CS} = Z_{\text{Couter}} + Z_{\text{CSinsul}} + Z_{\text{Sinner}} \quad (3.53)$$

$$Z_{L\_SA} = Z_{\text{Souter}} + Z_{\text{SAinsul}} + Z_{\text{Ainner}} \quad (3.54)$$

$$Z_{L\_AG} = Z_{\text{Aouter}} + Z_{\text{AGinsul}} + Z_{\text{earth}} \quad (3.55)$$

There is also mutual impedance associated to the core-screen loop (3.56) and to the screen-armour loop (3.57), both calculated directly using the equations previously explained.

$$Z_{Lm\_CS} = -Z_{\text{Smutual}} \quad (3.56)$$

$$Z_{Lm\_SA} = -Z_{\text{Amutual}} \quad (3.57)$$

We now have enough information to write the matrix equations describing the loop system for a three-phase system (3.58), where Z_{gmij} (or $Z_{\text{earth\_mutual}}$) is the mutual ground-return impedance between the armours of two phases.

$$
\begin{bmatrix} V_{CS1} \\ V_{SA1} \\ V_{AG1} \\ V_{CS2} \\ V_{SA2} \\ V_{AG2} \\ V_{CS3} \\ V_{SA3} \\ V_{AG3} \end{bmatrix} =
\begin{bmatrix}
Z_{L\_CS} & Z_{Lm\_CS} & 0 & 0 & 0 & 0 & 0 & 0 & 0 \\
Z_{Lm\_CS} & Z_{L\_SA} & Z_{Lm\_SA} & 0 & 0 & 0 & 0 & 0 & 0 \\
0 & Z_{Lm\_SA} & Z_{L\_AG} & 0 & 0 & Z_{\text{gm}12} & 0 & 0 & Z_{\text{gm}13} \\
0 & 0 & 0 & Z_{L\_CS} & Z_{Lm\_CS} & 0 & 0 & 0 & 0 \\
0 & 0 & 0 & Z_{Lm\_CS} & Z_{L\_SA} & Z_{Lm\_SA} & 0 & 0 & 0 \\
0 & 0 & Z_{\text{gm}12} & 0 & Z_{Lm\_SA} & Z_{L\_AG} & 0 & 0 & Z_{\text{gm}23} \\
0 & 0 & 0 & 0 & 0 & 0 & Z_{L\_CS} & Z_{Lm\_CS} & 0 \\
0 & 0 & 0 & 0 & 0 & 0 & Z_{Lm\_CS} & Z_{L\_SA} & Z_{Lm\_SA} \\
0 & 0 & Z_{\text{gm}13} & 0 & 0 & Z_{\text{gm}23} & 0 & Z_{Lm\_SA} & Z_{L\_AG}
\end{bmatrix}
\cdot
\begin{bmatrix} I_{CS1} \\ I_{SA1} \\ I_{AG1} \\ I_{CS2} \\ I_{SA2} \\ I_{AG2} \\ I_{CS3} \\ I_{SA3} \\ I_{AG3} \end{bmatrix}
$$

$$(3.58)$$

As useful as the loop impedance matrix is for understanding the currents circulating in a cable, the matrix that we need to calculate is the *series impedance matrix*, which can be used when analysing a more complex system. Equation (3.59) shows the relation between the series and loop impedance matrices, where the transformation matrix A is given by (3.60), for the cable previously described.

The transformation matrix A changes in function of the cable system, but the construction process is rather systematic. The first row of the matrix is correspondent to the core of the cable in Fig. 3.3. The only current in the core is the one from the core-screen loop. Thus, the first entry is 1 and the remaining 0. The second row is correspondent to the screen, thus, the first entry is -1, from the core-screen loop and the second entry is 1, from the screen-armour loop, whereas the remaining entries are 0. The same reasoning is then applied to the other rows of the transformation matrix. The matrix changes for pipe-type cables, but more on this subject later on.

$$[Z] = [A]^{-T}[Z_L][A]^{-1} \tag{3.59}$$

$$[A] = \begin{bmatrix} 1 & 0 & 0 & 0 & 0 & 0 & 0 & 0 & 0 \\ -1 & 1 & 0 & 0 & 0 & 0 & 0 & 0 & 0 \\ 0 & -1 & 1 & 0 & 0 & 0 & 0 & 0 & 0 \\ 0 & 0 & 0 & 1 & 0 & 0 & 0 & 0 & 0 \\ 0 & 0 & 0 & -1 & 1 & 0 & 0 & 0 & 0 \\ 0 & 0 & 0 & 0 & -1 & 1 & 0 & 0 & 0 \\ 0 & 0 & 0 & 0 & 0 & 0 & 1 & 0 & 0 \\ 0 & 0 & 0 & 0 & 0 & 0 & -1 & 1 & 0 \\ 0 & 0 & 0 & 0 & 0 & 0 & 0 & -1 & 1 \end{bmatrix} \tag{3.60}$$

Applying the relation (3.59) to (3.58), it is obtained the series impedance matrix (3.61).

$$[Z] = \begin{bmatrix} Z_{11} & Z_{12} & Z_{13} & Z_{gm12} & Z_{gm12} & Z_{gm12} & Z_{gm13} & Z_{gm13} & Z_{gm13} \\ Z_{12} & Z_{22} & Z_{23} & Z_{gm12} & Z_{gm12} & Z_{gm12} & Z_{gm13} & Z_{gm13} & Z_{gm13} \\ Z_{13} & Z_{23} & Z_{33} & Z_{gm12} & Z_{gm12} & Z_{gm12} & Z_{gm13} & Z_{gm13} & Z_{gm13} \\ Z_{gm12} & Z_{gm12} & Z_{gm12} & Z_{11} & Z_{12} & Z_{13} & Z_{gm23} & Z_{gm23} & Z_{gm23} \\ Z_{gm12} & Z_{gm12} & Z_{gm12} & Z_{12} & Z_{22} & Z_{23} & Z_{gm23} & Z_{gm23} & Z_{gm23} \\ Z_{gm12} & Z_{gm12} & Z_{gm12} & Z_{13} & Z_{23} & Z_{33} & Z_{gm23} & Z_{gm23} & Z_{gm23} \\ Z_{gm13} & Z_{gm13} & Z_{gm13} & Z_{gm23} & Z_{gm23} & Z_{gm23} & Z_{11} & Z_{12} & Z_{13} \\ Z_{gm13} & Z_{gm13} & Z_{gm13} & Z_{gm23} & Z_{gm23} & Z_{gm23} & Z_{12} & Z_{22} & Z_{23} \\ Z_{gm13} & Z_{gm13} & Z_{gm13} & Z_{gm23} & Z_{gm23} & Z_{gm23} & Z_{13} & Z_{23} & Z_{33} \end{bmatrix} \tag{3.61}$$

where:

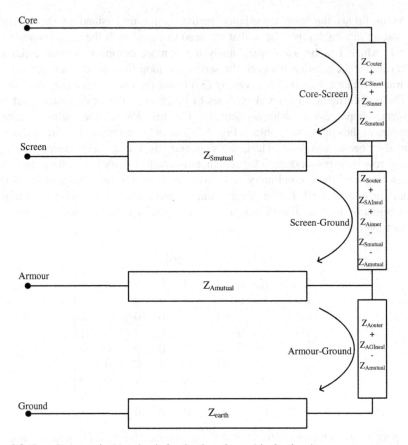

Fig. 3.3 Impedance equivalent circuit for the three loops (single-phase)

$Z_{11} = Z_{Couter} + Z_{CSinsul} + Z_{Sinner} + Z_{Souter} + Z_{SAinsul} + Z_{Ainner} + Z_{Aouter} + Z_{GAinsul} + Z_{earth} - 2Z_{Smutual} - 2Z_{Amutual}$

$Z_{22} = Z_{Souter} + Z_{SAinsul} + Z_{Aouter} + Z_{Ainner} + Z_{AGinsul} + Z_{earth} - 2Z_{Amutual}$

$Z_{33} = Z_{Ainner} + Z_{AGinsul} + Z_{earth}$

$Z_{12} = Z_{Souter} + Z_{SAinsul} + Z_{Aouter} + Z_{Ainner} + Z_{AGinsul} + Z_{earth} - Z_{Smutual} - 2Z_{Amutual}$

$Z_{13} = Z_{Aouter} + Z_{AGinsul} + Z_{earth} - Z_{Amutual}$

$Z_{23} = Z_{Aouter} + Z_{AGinsul} + Z_{earth}$

$Z_{gmij} = Z_{earth\_mutual}$

Another method that can be applied for single-core cables is to observe the loops to ground in Fig. 3.3 and write the series impedance matrix directly.

The first entry is the self-impedance of the core-ground loop (Z_{11}) and it can be obtained using the loops previously described. However, we should keep in mind that the voltage drop in the loop previously described is between the core and the screen, whereas it is between the core and the ground in the series impedance matrix. Therefore, the impedance of the core-ground loop is equal to (3.62).

$$Z_{11} = Z_{Couter} + Z_{CSinsul} + Z_{Sinner} + Z_{Souter} + Z_{SAinsul} + Z_{Ainner}$$
$$+ Z_{Aouter} + Z_{GAinsul} + Z_{earth} - 2Z_{Smutual} - 2Z_{Amutual} \tag{3.62}$$

The self-impedance of the screen-ground (Z_{22}) is also obtained using Fig. 3.3 and equal to (3.63), whereas the self-impedance of the armour-ground (Z_{33}) is equal to (3.64).

$$Z_{22} = Z_{Souter} + Z_{SAinsul} + Z_{Aouter} + Z_{Ainner} + Z_{AGinsul} + Z_{earth} - 2Z_{Amutual} \tag{3.63}$$

$$Z_{33} = Z_{Ainner} + Z_{AGinsul} + Z_{earth} \tag{3.64}$$

The mutual impedance between the loops core-ground and screen-ground (Z_{12}) is given by the part of the impedance that is common to both the core- and screen-ground loops (3.65). The mutual impedance between the other two loops, core armour (3.66) and screen armour (3.67), is calculated the same way

$$Z_{12} = Z_{Souter} + Z_{SAinsul} + Z_{Aouter} + Z_{Ainner} + Z_{AGinsul} + Z_{earth} - Z_{Smutual} - 2Z_{Amutual} \tag{3.65}$$

$$Z_{13} = Z_{Ainner} + Z_{AGinsul} + Z_{earth} - Z_{Amutual} \tag{3.66}$$

$$Z_{23} = Z_{Ainner} + Z_{AGinsul} + Z_{earth} \tag{3.67}$$

Finally, the ground mutual impedance between phases (Z_{gmij}) is simply equal to the mutual earth impedance (3.52), whereas the distance changes in function of the phases.

For armourless cables, the matrix (3.61) is reduced to a 6×6 by elimination of the rows and columns associated to the armour layers.

There are also changes in the formulas used to calculate the value of the entries. The components related to the armour (Z_{Ainner}, Z_{Aouter}, $Z_{Amutual}$, etc...) are eliminated and the $Z_{SAinsul}$ term becomes $Z_{SGinsul}$, but the matrix continues to be calculated in the same way.

Shunt Admittance Matrix
The shunt admittance is obtained in a similar way, by replacing (3.61), the impedance entries by the shunt admittance.

However, the admittance matrix is simpler than the impedance matrix, because of the several simplifications that can be applied. The screens of an HVAC cable are typically grounded in both ends, and also in middle points if cross-bond (see Sect. 1.2). Thus, it is safe to assume that the screen potential is virtually zero all along the cable.[1] As a result, the electric field is confined to each of the phases and there is no shunt admittance between different phases. Figure 3.4 shows a visual description of the electric field distribution and the equivalent capacitance circuit.

---

[1] In reality, there is a small voltage in the screen.

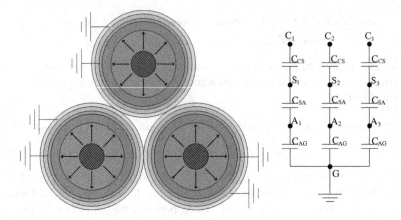

Fig. 3.4 Electric field in three single-core cables installed in a trefoil configuration and equivalent capacitance circuit

As a result, the matrix may be simplified and several of the matrix entries made zero (3.68). Notice that this reasoning is only for cables where both ends are grounded.

$$
[Y] = \begin{bmatrix}
Y_{C1C1} & Y_{C1S1} & 0 & 0 & 0 & 0 & 0 & 0 & 0 \\
Y_{S1C1} & Y_{S1S1} & Y_{S1A1} & 0 & 0 & 0 & 0 & 0 & 0 \\
0 & Y_{A1S1} & Y_{A1A1} & 0 & 0 & 0 & 0 & 0 & 0 \\
0 & 0 & 0 & Y_{C2C2} & Y_{C2S2} & 0 & 0 & 0 & 0 \\
0 & 0 & 0 & Y_{S2C2} & Y_{S2S2} & Y_{S2A2} & 0 & 0 & 0 \\
0 & 0 & 0 & 0 & Y_{A2S2} & Y_{A2A2} & 0 & 0 & 0 \\
0 & 0 & 0 & 0 & 0 & 0 & Y_{C3C3} & Y_{C3S3} & 0 \\
0 & 0 & 0 & 0 & 0 & 0 & Y_{S3C3} & Y_{S3S3} & Y_{S3A3} \\
0 & 0 & 0 & 0 & 0 & 0 & 0 & Y_{A3S3} & Y_{A3A3}
\end{bmatrix}
$$

(3.68)

A shunt admittance is described by the typical expression (3.69), where the real part is normally considered zero.[2]

$$
Y_i = G_i + j\omega C_i \tag{3.69}
$$

By applying (3.69) in the equivalent capacitance circuit of Fig. 3.4, the entries of the shunt admittance matrix (3.70) are obtained, which can be rewritten as (3.71).

---

[2] There are also dielectric losses that could be incorporated into the matrix, but they are small when compared with the capacitance.

$$Y_{CiCi} = j\omega C_{CS}$$
$$Y_{SiSi} = j\omega C_{CS} + j\omega C_{SA}$$
$$Y_{AiAi} = j\omega C_{SA} + j\omega C_{AG} \tag{3.70}$$
$$Y_{CiSi} = -j\omega C_{CS}$$
$$Y_{SiAi} = -j\omega C_{SA}$$

$$[Y] = \begin{bmatrix}
j\omega C_{CS} & -j\omega C_{CS} & 0 & 0 & 0 & 0 & 0 & 0 & 0 \\
-j\omega C_{CS} & j\omega(C_{CS}+C_{SA}) & -j\omega C_{SA} & 0 & 0 & 0 & 0 & 0 & 0 \\
0 & -j\omega C_{SA} & j\omega(C_{SA}+C_{SG}) & 0 & 0 & 0 & 0 & 0 & 0 \\
0 & 0 & 0 & j\omega C_{CS} & -j\omega C_{CS} & 0 & 0 & 0 & 0 \\
0 & 0 & 0 & -j\omega C_{CS} & j\omega(C_{CS}+C_{SA}) & -j\omega C_{SA} & 0 & 0 & 0 \\
0 & 0 & 0 & 0 & -j\omega C_{SA} & j\omega(C_{SA}+C_{SG}) & 0 & 0 & 0 \\
0 & 0 & 0 & 0 & 0 & 0 & j\omega C_{CS} & -j\omega C_{CS} & 0 \\
0 & 0 & 0 & 0 & 0 & 0 & -j\omega C_{CS} & j\omega(C_{CS}+C_{SA}) & -j\omega C_{SA} \\
0 & 0 & 0 & 0 & 0 & 0 & 0 & -j\omega C_{SA} & j\omega(C_{SA}+C_{SG})
\end{bmatrix}$$
$$\tag{3.71}$$

Similar to the impedance matrix, the admittance matrix is also reduced to a 6×6 for an armourless cable. The rows and columns associated to the armours are eliminated and the value of Y_{SiSi} is given by $j\omega(C_{CS} + C_{SG})$.

3.3.1.1 Example

A Matlab code where the several parameters are calculated is available at the book's webpage.

It may be difficult to understand all the formulas given above when studying this subject for the first time. The example presented next shows how to calculate the impedance and admittance matrix for a three-phase single-core cable with core and screen. Table 3.1 shows the cable's data and the thickness of the several layers, whereas Fig. 3.5 shows a cross-section of the cable (armourless cable).

The core of the cable is made of compact stranded aluminium wires. It is therefore necessary to correct the resistivity of the core (3.72) as explained in (3.31).

Table 3.1 Cable data

Layer	Thickness (mm)	Material
Conductor	41.5*	Aluminium, round, compacted
Conductor screen	1.5	Semi-conductive PE
Insulation	17	Dry cured XLPE
Insulation screen	1	Semi-conductive PE
Longitudinal water barrier	0.6	Swelling tape
Copper wire screen	95**	Copper
Longitudinal water barrier	0.6	Swelling tape
Radial water barrier	0.2	Aluminium laminate
Outer cover	4	High-density PE
Complete cable	95*	–

* Diameter
** Cross-section

Fig. 3.5 Cross-section of a
150 kV single-core land
cable

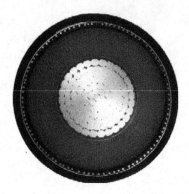

$$\rho'_C = \rho_C \frac{\pi R_1}{A_C} \Leftrightarrow \rho' = 2.826 \times 10^{-8} \frac{\pi \cdot 20.75^2}{1200} \Leftrightarrow \rho' = 3.186 \times 10^{-8}\ \Omega m \quad (3.72)$$

The resistivity of the screen also needs to be corrected. The expression is more complicated than for the conductor for two reasons. First, it is necessary to correct the resistivity of the copper wires (3.73) and second the total resistivity of the screen is given by the parallel of the copper wires and the aluminium foil (3.74).

$$\rho'_{S,Cu} = \rho_{S,Cu} \frac{\pi\left(R^2_{\text{Wires}} - R^2_2\right)}{A_S} \Leftrightarrow \rho'_{S,Cu} = 1.724 \times 10^{-8} \frac{\pi(41.96^2 - 40.85^2)}{95}$$
$$\Leftrightarrow \rho'_{S,Cu} = 5.240 \times 10^{-8}\ \Omega m$$

$$(3.73)$$

$$\rho'_S = \frac{\rho_{Cu}\rho_{Al}(A_{Cu} + A_{Al})}{\rho_{Cu}A_{Al} + \rho_{Al}A_{Cu}} \Leftrightarrow \rho'_S = \frac{5.240 \times 10^{-8} \cdot 2.826 \times 10^{-8}(288.8 \times 10^{-6} + 212.9 \times 10^{-6})}{5.240 \times 10^{-8} \cdot 212.9 \times 10^{-6} + 2.826 \times 10^{-8} \cdot 288.8 \times 10^{-6}}$$
$$\Leftrightarrow \rho'_S = 8.98 \times 10^{-8}\ \Omega m$$

$$(3.74)$$

The reciprocal of the complex penetration depths for the conductor and screen is given by respectively (3.75) and (3.76).

$$m_C = \sqrt{\frac{j\omega\mu}{\rho_C}} \Leftrightarrow m_C = \sqrt{\frac{j\omega 4\pi \times 10^{-7}}{3.186 \times 10^{-8}}} \Leftrightarrow m_C = 78.71 + j78.71\ m^{-1} \quad (3.75)$$

$$m_S = \sqrt{\frac{j\omega\mu}{\rho_S}} \Leftrightarrow m_C = \sqrt{\frac{j\omega 4\pi \times 10^{-7}}{8.98 \times 10^{-8}}} \Leftrightarrow m_C = 46.88 + j46.88\ m^{-1} \quad (3.76)$$

The impedance of the several formulas used to write the series impedance matrix is given next:

$$Z_{\text{Couter}} = 2.6687 \times 10^{-5} + j1.468 \times 10^{-5}\,\Omega/\text{m}$$
$$Z_{\text{CSinsul}} = 0 + j4.256 \times 10^{-5}\,\Omega/\text{m}$$
$$Z_{\text{Sinner}} = 1.790 \times 10^{-4} + j9.791 \times 10^{-7}\,\Omega/\text{m}$$
$$Z_{\text{Souter}} = 1.7790 \times 10^{-4} + j9.353 \times 10^{-7}\,\Omega/\text{m}$$
$$Z_{\text{SGinsul}} = 0 + j6.605 \times 10^{-6}\,\Omega/\text{m}$$
$$Z_{\text{Smutual}} = 1.790 \times 10^{-4} - j4.784 \times 10^{-8}\,\Omega/\text{m}$$
$$Z_{\text{earth}} \approx 4.945 \times 10^{-5} + j6.209 \times 10^{-4}\,\Omega/\text{m}*$$
$$Z_{\text{earth\_mutual}} \approx 4.945 \times 10^{-5} + j5.774 \times 10^{-4}\,\Omega/\text{m}*$$

*There is a small variation for each of the phases
The series impedance is given by (3.77).

$$[Z] = \begin{bmatrix} 0.0761 + j0.688 & 0.0494 + j0.629 & 0.0494 + j0.577 & 0.0494 + j0.577 & 0.0494 + j0.577 & 0.0494 + j0.577 \\ 0.0494 + j0.629 & 0.2284 + j0.628 & 0.0494 + j0.577 & 0.0494 + j0.577 & 0.0494 + j0.577 & 0.0494 + j0.577 \\ 0.0494 + j0.577 & 0.0494 + j0.577 & 0.0761 + j0.688 & 0.0494 + j0.629 & 0.0494 + j0.577 & 0.0494 + j0.577 \\ 0.0494 + j0.577 & 0.0494 + j0.577 & 0.0494 + j0.629 & 0.2284 + j0.628 & 0.0494 + j0.577 & 0.0494 + j0.577 \\ 0.0494 + j0.577 & 0.0494 + j0.577 & 0.0494 + j0.577 & 0.0494 + j0.577 & 0.0761 + j0.688 & 0.0494 + j0.629 \\ 0.0494 + j0.577 & 0.0494 + j0.577 & 0.0494 + j0.577 & 0.0494 + j0.577 & 0.0494 + j0.629 & 0.2284 + j0.628 \end{bmatrix} \text{m}\Omega/\text{m}$$

$$(3.77)$$

The series impedance matrix is according to the expected:

- The matrix is symmetric.
- There are only four different entries:

  - Core self-impedance: $0.0761 + j0.668$
  - Screen self-impedance: $0.2284 + j0.628$
  - Core-screen mutual impedance (same phase): $0.0494 + j0.629$
  - Mutual impedance between phases: $0.0494 + j0.577$.
- The self-impedances are larger than the mutual impedances.
- The mutual-impedance in the same phase is larger than the mutual impedance between phases and the difference is in the imaginary part of the equation (inductive coupling).

The admittance matrix (3.81) is easier to calculate. The permittivity of the insulation is corrected to include the semi-conductive layers (3.78), whereas the two entries of the admittance matrix are given by (3.79) and (3.80).

$$\varepsilon' = \varepsilon \frac{\ln\left(\frac{R_2}{R_1}\right)}{\ln\left(\frac{b}{a}\right)} \Leftrightarrow \varepsilon' = 2.5\frac{\ln\left(\frac{40.85}{20.75}\right)}{\ln\left(\frac{39.50}{22.25}\right)} \Leftrightarrow \varepsilon' = 2.95 \qquad (3.78)$$

$$Y_{CiCi} = j\omega C_{CS} \Leftrightarrow Y_{CiCi} = 7.612 \times 10^{-8}\,\text{S/m} \qquad (3.79)$$

$$Y_{SiSi} = j\omega C_{CS} + j\omega C_{SA} \Leftrightarrow Y_{SiSi} = 7.612 \times 10^{-8} + 3.834 \times 10^{-7}$$
$$\Leftrightarrow Y_{SiSi} = 4.595 \times 10^{-7}\,\text{S/m} \qquad (3.80)$$

$$[Y] = \begin{bmatrix} j0.0761 & -j0.0761 & 0 & 0 & 0 & 0 \\ -j0.0761 & j0.4595 & 0 & 0 & 0 & 0 \\ 0 & 0 & j0.0761 & -j0.0761 & 0 & 0 \\ 0 & 0 & -j0.0761 & j0.4595 & 0 & 0 \\ 0 & 0 & 0 & 0 & j0.0761 & -j0.0761 \\ 0 & 0 & 0 & 0 & -j0.0761 & j0.4595 \end{bmatrix} \mu S/m$$

$$(3.81)$$

3.3.2 Cross-Bonded Cable

The screen of a cross-bond cable is cross-connected as explained in Sect. 1.2 and shown in Fig. 3.6. This arrangement changes the *series impedance matrix*, because of the changes in the inductive coupling between phases. As a result, it is necessary to write three matrices, one for each minor section.

The matrix describing the impedance of the first minor section is equal to the one describing a cable bonded in both ends (3.28), whereas the matrices describing the other two minor sections are like shown in (3.82) and (3.83). The changes in the two matrices are related to the changes in the position of the screens, whereas the coupling between the cores, and between armours, remains unchanged. As an example, in the matrix describing the first minor section, the second column is the screen of Phase A, whereas in the second minor section, it is correspondent to the screen of Phase B and to Phase C in the third minor section. However, the first column describes the core of Phase A for all three matrices.

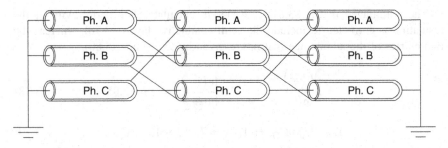

Fig. 3.6 Major-section of a cross-bond cable

$$[Z_{S2}] = \begin{array}{c} \begin{array}{ccccccccc} \text{Core1} & \text{Screen2} & \text{Arm.1} & \text{Core2} & \text{Screen3} & \text{Arm.2} & \text{Core3} & \text{Screen 1} & \text{Arm.3} \end{array} \\ \begin{bmatrix} Z_{C1C1} & Z_{C1S2} & Z_{C1A1} & Z_{C1C2} & Z_{C1S3} & Z_{C1A2} & Z_{C1C3} & Z_{C1S1} & Z_{C1A3} \\ Z_{S2C1} & Z_{S2S2} & Z_{S2A1} & Z_{S2C2} & Z_{S2S3} & Z_{S2A2} & Z_{S2C3} & Z_{S2S1} & Z_{S2A3} \\ Z_{A1C1} & Z_{A1S2} & Z_{A1A1} & Z_{A1C2} & Z_{A1S3} & Z_{A1A2} & Z_{A1C3} & Z_{A1S1} & Z_{A1A3} \\ Z_{C2C1} & Z_{C2S2} & Z_{C2A1} & Z_{C2C2} & Z_{C2S3} & Z_{C2A2} & Z_{C2C3} & Z_{C2S1} & Z_{C2A3} \\ Z_{S3C1} & Z_{S3S2} & Z_{S3A1} & Z_{S3C2} & Z_{S3S3} & Z_{S3A2} & Z_{S3C3} & Z_{S3S1} & Z_{S3A3} \\ Z_{A2C1} & Z_{A2S2} & Z_{A2A1} & Z_{A2C2} & Z_{A2S3} & Z_{A2A2} & Z_{A2C3} & Z_{A2S1} & Z_{A2A3} \\ Z_{C3C1} & Z_{C3S2} & Z_{C3A1} & Z_{C3C2} & Z_{C3S3} & Z_{C3A2} & Z_{C3C3} & Z_{C3S1} & Z_{C3A3} \\ Z_{S1C1} & Z_{S1S2} & Z_{S1A1} & Z_{S1C2} & Z_{S1S3} & Z_{S1A2} & Z_{S1C3} & Z_{S1S1} & Z_{S1A3} \\ Z_{A3C1} & Z_{A3S2} & Z_{A3A1} & Z_{A3C2} & Z_{A3S3} & Z_{A3A2} & Z_{A3C3} & Z_{A3S1} & Z_{A3A3} \end{bmatrix} \end{array}$$

$$(3.82)$$

$$[Z_{S3}] = \begin{array}{c} \begin{array}{ccccccccc} \text{Core1} & \text{Screen3} & \text{Arm.1} & \text{Core 2} & \text{Screen1} & \text{Arm.2} & \text{Core3} & \text{Screen 2} & \text{Arm.3} \end{array} \\ \begin{bmatrix} Z_{C1C1} & Z_{C1S3} & Z_{C1A1} & Z_{C1C2} & Z_{C1S1} & Z_{C1A2} & Z_{C1C3} & Z_{C1S2} & Z_{C1A3} \\ Z_{S3C1} & Z_{S3S3} & Z_{S3A1} & Z_{S3C2} & Z_{S3S1} & Z_{S3A2} & Z_{S3C3} & Z_{S3S2} & Z_{S3A3} \\ Z_{A1C1} & Z_{A1S3} & Z_{A1A1} & Z_{A1C2} & Z_{A1S1} & Z_{A1A2} & Z_{A1C3} & Z_{A1S2} & Z_{A1A3} \\ Z_{C2C1} & Z_{C2S3} & Z_{C2A1} & Z_{C2C2} & Z_{C2S1} & Z_{C2A2} & Z_{C2C3} & Z_{C2S2} & Z_{C2A3} \\ Z_{S1C1} & Z_{S1S3} & Z_{S1A1} & Z_{S1C2} & Z_{S1S1} & Z_{S1A2} & Z_{S1C3} & Z_{S1S2} & Z_{S1A3} \\ Z_{A2C1} & Z_{A2S3} & Z_{A2A1} & Z_{A2C2} & Z_{A2S1} & Z_{A2A2} & Z_{A2C3} & Z_{A2S2} & Z_{A2A3} \\ Z_{C3C1} & Z_{C3S3} & Z_{C3A1} & Z_{C3C2} & Z_{C3S1} & Z_{C3A2} & Z_{C3C3} & Z_{C3S2} & Z_{C3A3} \\ Z_{S2C1} & Z_{S2S3} & Z_{S2A1} & Z_{S2C2} & Z_{S2S1} & Z_{S2A2} & Z_{S2C3} & Z_{S2S2} & Z_{S2A3} \\ Z_{A3C1} & Z_{A3S3} & Z_{A3A1} & Z_{A3C2} & Z_{A3S1} & Z_{A3A2} & Z_{A3C3} & Z_{A3S2} & Z_{A3A3} \end{bmatrix} \end{array}$$

$$(3.83)$$

The total is then given by averaging the three minor sections (3.84). This result is only valid assuming an ideal cross-bonding, i.e. a perfect transposition of the screens. We will see in Sect. 3.5 that if the cable has few cross-bond sections, (3.84) is no longer valid.

$$[Z] = \frac{[Z_{S1}] + [Z_{S2}] + [Z_{S3}]}{3} \tag{3.84}$$

A similar reasoning could be applied to the admittance matrix, but it is not necessary as the values are identical for all the phases.

3.3.3 Three-Core Cables (Pipe-Type)

Figure 3.7 shows a typical pipe-type cable whereupon we are going to apply the loop analysis. Figure 3.8 shows the two loops associated to a pipe-type cable. For pipe-type cables with individual armours for each phase, a third loop would be added between the screen and the pipe, similar to Fig. 3.3.

It is considered in this model (Fig. 3.8) that the pipe thickness is larger than the penetration depth into the pipe, something that is usually true except for low frequencies. In other words, the pipe thickness is considered to be infinite.

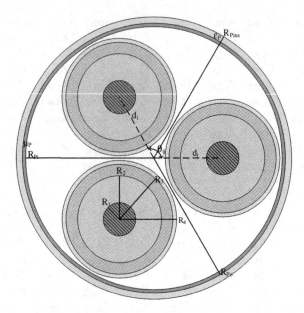

Fig. 3.7 Cross-section of a pipe-type cable

Consequently, it can be assumed that all the current returns in the pipe and that the ground may be ignored.

We know from the single-core example that the conductor self-impedance can be calculated by the core-ground loop of Fig. 3.8, which is easier than the formal method.

The core-screen loop is identical to the one of a single-core cable, as there are no changes in neither the core, nor the insulation and the screen.

The parameters in the screen-pipe loop depend on the pipe and configuration of the cables inside the pipe.

The *insulation series impedance* is given by (3.85), which can be divided into two parts. The first part refers to the insulation of each screen, layer between R_3 and R_4 in Fig. 3.7 and the calculation is identical to the one of a single-core cable. Yet, three-core cables do not always have an individual insulation around each screen, in these cases, R_4 is equal to R_3 and the final result is zero.

The second part of the equation refers to pipe inner insulation and the formula is changed to accommodate the non-concentricity of the cable with respect to the centre of the pipe. For example, if the pipe-type cable had only one conductor in a central position, the variable d would be zero and the second part of the equation would have the same structure as the first part. Notice that the value of Z_{SPinsul} is only equal for the three phases if each phase is at the same distance from the centre of the pipe. Thus, if the pipe-type cable has the conductors in the bottom of the pipe, the insulation series impedance will not be the same for all three phases.

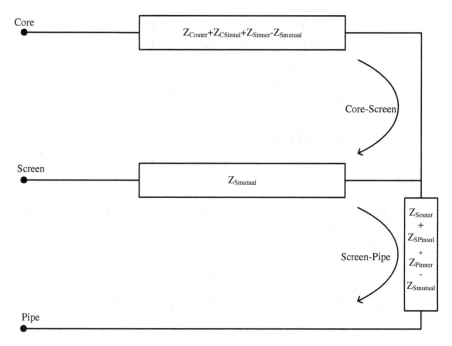

Fig. 3.8 Impedance equivalent circuit for the two loops (pipe-type with infinite pipe thickness)

$$Z_{SPinsul} = \frac{j\omega\mu_{out\_ins}}{2\pi}\ln\left(\frac{R_4}{R_3}\right) + \frac{j\omega\mu_0\mu_P}{2\pi}\ln\left(\frac{R_{Pi}}{R_4}\left(1 - \left(\frac{d}{R_{Pi}}\right)^2\right)\right) \quad (3.85)$$

The *inner series impedance* of the pipe is given by (3.86) and it is also subdivided into two parts. The first part of the formula is similar to the expression used to calculate the inner series impedance of a screen, but with the parameters changed for the pipe. The second part of the equation is present because of the non-concentricity of the cable.

$$Z_{Pinner} = \frac{j\omega\mu_0}{2\pi}\left[\frac{\mu_p}{m_pR_{Pi}}\frac{J_o(m_pR_{Pi})K_1(m_pR_{Po}) + K_o(m_pR_{Pi})J_1(m_pR_{Po})}{K_1(m_pR_{Pi})J_1(m_pR_{Po}) - J_1(m_pR_{Pi})K_1(m_pR_{Po})}\right.$$
$$\left. + 2\mu_p\sum_{n=1}^{\infty}\left(\frac{d}{R_{Pi}}\right)^{2n}\frac{1}{n(1+\mu_P) + m_pR_{Pi}\frac{K_{n-1}(m_pR_{Pi})}{K_n(m_pR_{Pi})}}\right] \quad (3.86)$$

The impedance matrix also contains the *mutual impedance* between the cables inside the pipe, which is calculated using (3.87).

$$Z_{Pipe\_mutual} = \frac{j\omega\mu_0}{2\pi} \left[\ln\left(\frac{R_{Pi}}{\sqrt{d_i^2 + d_j^2 - 2d_i d_j \cos(\theta_{ij})}} \right) \right.$$

$$+ \frac{\mu_p}{m_p R_{Pi}} \frac{J_o(m_p R_{Pi})K_1(m_p R_{Po}) + K_o(m_p R_{Pi})J_1(m_p R_{Po})}{K_1(m_p R_{Pi})J_1(m_p R_{Po}) - J_1(m_p R_{Pi})K_1(m_p R_{Po})}$$

$$\left. + \sum_{n=1}^{\infty} \left(\frac{d_i d_j}{R_{Pi}^2} \right)^n \cos(n\theta_{ij}) \left(\frac{2}{n(1+\mu_P) + m_p R_{Pi} \frac{K_{n-1}(m_p R_{Pi})}{K_n(m_p R_{Pi})}} - \frac{1}{n} \right) \right]$$

$$(3.87)$$

At this point, we have all the information necessary to write the *impedance matrix*, which is given by (3.88).

$$[Z] = \begin{matrix} & \text{Core 1} & \text{Screen 1} & \text{Core 2} & \text{Screen 2} & \text{Core 3} & \text{Screen 3} \\ & \begin{pmatrix} Z_{11} & Z_{12} & Z_{pm12} & Z_{pm12} & Z_{pm13} & Z_{pm13} \\ Z_{12} & Z_{22} & Z_{pm12} & Z_{pm12} & Z_{pm13} & Z_{pm13} \\ Z_{pm12} & Z_{pm12} & Z_{11} & Z_{12} & Z_{pm23} & Z_{pm23} \\ Z_{pm12} & Z_{pm12} & Z_{12} & Z_{22} & Z_{pm23} & Z_{pm23} \\ Z_{pm13} & Z_{pm13} & Z_{pm23} & Z_{pm23} & Z_{11} & Z_{12} \\ Z_{pm13} & Z_{pm13} & Z_{pm23} & Z_{pm23} & Z_{12} & Z_{22} \end{pmatrix} \end{matrix} \quad (3.88)$$

where:

$$Z_{11} = Z_{Couter} + Z_{CSinsul} + Z_{Sinner} + Z_{Souter} + Z_{SPinsul} + Z_{Pinner} - 2Z_{Smutual}$$

$$Z_{22} = Z_{Souter} + Z_{SPinsul} + Z_{Pinner}$$

$$Z_{12} = Z_{Souter} + Z_{SPinsul} + Z_{Pinner} - Z_{Smutual}$$

$$Z_{pmij} = Z_{Pipe\_mutual}$$

The comparison of the pipe-type cable and single-core cable impedance matrices shows that they are very similar. The similarity makes sense, as it was assumed in the model that ground return is substituted by a "pipe return". Thus, the only differences are in the substitution of the Z_{earth} by Z_{Pinner} and of the Z_{earth_mutual} by Z_{Pipe_mutual}.

We have previously assumed that the pipe thickness was larger than the skin depth and that no current would return through the ground. This assumption is not true for low frequencies and in those situations, it is necessary to consider the ground. Figure 3.9 shows the loops for this situation.

There are three new parameters that were present before: $Z_{Pmutual}$, Z_{Pouter} and $Z_{PGinsul}$.

The formula used to calculate the *mutual series impedance* for the pipe is similar to the one used before for the screen and armour, by substituting the variables to the ones associated to the pipe as shown in (3.89).

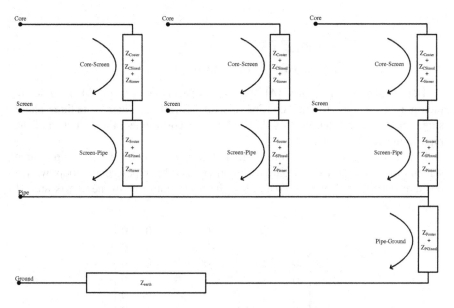

Fig. 3.9 Impedance equivalent circuit for the two loops (pipe type with finite pipe thickness)

$$Z_{\mathrm{Pmutual}} = \frac{\rho_p}{2\pi R_{Pi} R_{Po}} \frac{1}{K_1\left(m_p R_{Pi}\right) J_1\left(m_p R_{Po}\right) - J_1\left(m_p R_{Pi}\right) K_1\left(m_p R_{Po}\right)} \tag{3.89}$$

Lastly we have the pipe-ground loop, whose formulas have nothing new when compared with the ones previously used. The pipe *outer series impedance* is given by (3.90), whereas the *insulation series impedance* of the pipe is given by (3.91), which does not exist if the pipe does not have an insulation.

$$Z_{\mathrm{Pouter}}(\omega) = \frac{\rho_p m_p}{2\pi R_{Po}} \coth\left(m_p(R_{Po} - R_{Pi})\right) + \frac{\rho_p}{2\pi R_{Po}(R_{Pi} + R_{Po})} \tag{3.90}$$

$$Z_{\mathrm{PGinsul}}(\omega) = \frac{j\omega\mu_p}{2\pi} \ln\left(\frac{R_{\mathrm{Pins}}}{R_{Po}}\right) \tag{3.91}$$

The loop equations are written as (3.92).

$$
\begin{bmatrix} V_{CS1} \\ V_{SP1} \\ V_{CS1} \\ V_{SP1} \\ V_{CS1} \\ V_{SP1} \\ V_{PG} \end{bmatrix} = \begin{bmatrix} Z_{L\_CS} & Z_{Lm\_CS} & 0 & 0 & 0 & 0 & 0 \\ Z_{Lm\_CS} & Z_{L\_SP} & 0 & Z_{\mathrm{pm}12} & 0 & Z_{\mathrm{pm}13} & Z_{\mathrm{pm}1g} \\ 0 & 0 & Z_{L\_CS} & Z_{Lm\_CS} & 0 & 0 & 0 \\ 0 & Z_{\mathrm{pm}12} & Z_{Lm\_CS} & Z_{L\_SP} & 0 & Z_{\mathrm{pm}23} & Z_{\mathrm{pm}2g} \\ 0 & 0 & 0 & 0 & Z_{L\_CS} & Z_{Lm\_CS} & 0 \\ 0 & Z_{\mathrm{pm}13} & 0 & Z_{\mathrm{pm}23} & Z_{Lm\_CS} & Z_{L\_SP} & Z_{\mathrm{pm}3g} \\ 0 & Z_{\mathrm{pm}1g} & 0 & Z_{\mathrm{pm}2g} & 0 & Z_{\mathrm{pm}3g} & Z_{L\_PG} \end{bmatrix} \cdot \begin{bmatrix} I_{CS1} \\ I_{SP1} \\ I_{CS1} \\ I_{SP1} \\ I_{CS1} \\ I_{SP1} \\ I_{PG} \end{bmatrix}
$$

$$\tag{3.92}$$

where:

$$Z_{pmig} = -Z_{Pmutual}$$
$$Z_{L\_PG} = Z_{Pouter} + Z_{PGinsul} + Z_{earth}$$

The transformation matrix $[A]$ also has to be changed to (3.93), in order to incorporate the pipe-ground loop. The first six rows are identical to the three-phase single-core case, but the last row has to be changed to reflect the influence of the pipe in the current's path.

Figure 3.9 shows that the four currents loops are associated to the pipe: three screen-pipe loops with a negative direction and the pipe-ground loop with a positive direction. Thus, the screen terms in the seventh row of the matrix are -1 and the pipe term is 1.

$$[A] = \begin{bmatrix} 1 & 0 & 0 & 0 & 0 & 0 & 0 \\ -1 & 1 & 0 & 0 & 0 & 0 & 0 \\ 0 & 0 & 1 & 0 & 0 & 0 & 0 \\ 0 & 0 & -1 & 1 & 0 & 0 & 0 \\ 0 & 0 & 0 & 0 & 1 & 0 & 0 \\ 0 & 0 & 0 & 0 & -1 & 1 & 0 \\ 0 & -1 & 0 & -1 & 0 & -1 & 1 \end{bmatrix} \tag{3.93}$$

Applying (3.59)–(3.92) and (3.93), (3.94) is obtained as the series impedance matrix.

$$
\begin{array}{cccccccc}
\text{Core 1} & \text{Screen 1} & \text{Core 2} & \text{Screen 2} & \text{Core 3} & \text{Screen 3} & \text{Pipe}
\end{array}
$$

$$[Z] = \begin{bmatrix} Z_{11} & Z_{12} & Z_{pm12} & Z_{pm12} & Z_{pm13} & Z_{pm13} & Z_{pm1p} \\ Z_{12} & Z_{22} & Z_{pm12} & Z_{pm12} & Z_{pm13} & Z_{pm13} & Z_{pm1p} \\ Z_{pm12} & Z_{pm12} & Z_{11} & Z_{12} & Z_{pm23} & Z_{pm23} & Z_{pm2p} \\ Z_{pm12} & Z_{pm12} & Z_{12} & Z_{22} & Z_{pm23} & Z_{pm23} & Z_{pm2p} \\ Z_{pm13} & Z_{pm13} & Z_{gm23} & Z_{gm23} & Z_{11} & Z_{12} & Z_{pm3p} \\ Z_{pm13} & Z_{pm13} & Z_{pm23} & Z_{pm23} & Z_{12} & Z_{22} & Z_{pm3p} \\ Z_{pm1p} & Z_{pm1p} & Z_{pm2p} & Z_{pm2p} & Z_{pm3p} & Z_{pm3p} & Z_{33} \end{bmatrix} \tag{3.94}$$

where:

$$Z_{11} = Z_{Couter} + Z_{CSinsul} + Z_{Sinner} + Z_{Souter} + Z_{SPinsul} + Z_{Pinner} + Z_{Pouter} + Z_{PGinsul} + Z_{earth} - 2Z_{Smutual} - 2Z_{Pmutual}$$

$$Z_{22} = Z_{Souter} + Z_{SPinsul} + Z_{pinner} + Z_{Pouter} + Z_{PGinsul} + Z_{earth} - 2Z_{Pmutual}$$

$$Z_{33} = Z_{Pouter} + Z_{PGinsul} + Z_{earth}$$

$$Z_{12} = Z_{Souter} + Z_{SPinsul} + Z_{Pinner} + Z_{Pouter} + Z_{PGinsul} + Z_{earth} - Z_{Smutual} - 2Z_{Pmutual}$$

$$Z_{pmij} = Z_{Pipe\_mutual} + Z_{Pouter} + Z_{PGinsul} + Z_{earth} - 2Z_{Pmutual}$$

$$Z_{pmip} = Z_{Pouter} + Z_{PGinsul} + Z_{earth} - Z_{Pmutual}$$

A first comparison of the two series impedance matrices seems to indicate that the thickness of the pipe has a big influence on the results. The matrix increases, the value of the entries is changed and the two matrices are completely different.

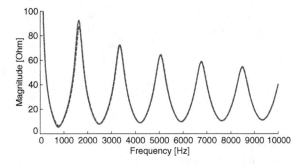

Fig. 3.10 Frequency spectrum of a 50 km cable. *Solid line* finite pipe thickness, *dashed line* infinite pipe thickness

However, a more detailed analysis of the results shows that the influence is in general minimal and that the current/voltage waveforms and amplitudes virtually do not change. Figure 3.10 shows that even the frequency spectrum for both cases is alike (both cases are for the cable studied in example 3.3.3.1). The similarity is expected, as the penetration depth is typically smaller than the cable thickness.

This is not a general rule and it is dependent on the geometry, but it is true for the most common configurations and layout conditions.

Admittance matrix
Like for the impedance matrix, the admittance matrix of a pipe-type cable is also more complex than that of a single-core cable. The total admittance is the combination of several contributions and for that reason, it is easier to write each contribution using potential coefficient matrices, which can be simply added and later invert the summation of matrix to obtain the admittance matrix (3.95).

$$[Y] = j\omega[P]^{-1} \qquad (3.95)$$

The matrix can be divided into three parts (3.96):

- The potential coefficients of the core and screen (P_i),
- The potential coefficients between the cables and the pipe (P_p),
- The potential coefficients between the pipe and the ground (P_c).

$$[P] = [P_i] + [P_p] + [P_c] \qquad (3.96)$$

The potential coefficient matrix of the core and screen (3.97) is identical to the potential coefficient matrix of a single-core cable. There is a potential between the core and the screen and between the screen and the ground, but none between phases.

$$[P_i] = \begin{bmatrix} P_{\text{cond}} + P_{\text{screen}} & P_{\text{screen}} & 0 & 0 & 0 & 0 & 0 \\ P_{\text{screen}} & P_{\text{screen}} & 0 & 0 & 0 & 0 & 0 \\ 0 & 0 & P_{\text{cond}} + P_{\text{screen}} & P_{\text{screen}} & 0 & 0 & 0 \\ 0 & 0 & P_{\text{screen}} & P_{\text{screen}} & 0 & 0 & 0 \\ 0 & 0 & 0 & 0 & P_{\text{cond}} + P_{\text{screen}} & P_{\text{screen}} & 0 \\ 0 & 0 & 0 & 0 & P_{\text{screen}} & P_{\text{screen}} & 0 \\ 0 & 0 & 0 & 0 & 0 & 0 & 0 \end{bmatrix}$$

(3.97)

The potential coefficients between the three phases and the pipe are given in (3.98). It is again necessary to take into account the non-concentricity of the cable.

$$[P_P] = \begin{bmatrix} P_{Pii} & P_{Pii} & P_{Pij} & P_{Pij} & P_{Pij} & P_{Pij} & 0 \\ P_{Pii} & P_{Pii} & P_{Pij} & P_{Pij} & P_{Pij} & P_{Pij} & 0 \\ P_{Pij} & P_{Pij} & P_{Pii} & P_{Pii} & P_{Pij} & P_{Pij} & 0 \\ P_{Pij} & P_{Pij} & P_{Pii} & P_{Pii} & P_{Pij} & P_{Pij} & 0 \\ P_{Pij} & P_{Pij} & P_{Pij} & P_{Pij} & P_{Pii} & P_{Pii} & 0 \\ P_{Pij} & P_{Pij} & P_{Pij} & P_{Pij} & P_{Pii} & P_{Pii} & 0 \\ 0 & 0 & 0 & 0 & 0 & 0 & 0 \end{bmatrix}$$

(3.98)

$$P_{Pii} = \frac{1}{2\pi\varepsilon_{\text{pipe}}} \ln\left(\frac{R_{Pi}}{R_4}\left(1 - \frac{d_i}{R_4}\right)^2\right)$$

(3.99)

$$P_{Pij} = \frac{1}{2\pi\varepsilon_{\text{pipe}}}\left(\ln\left(\frac{R_{Pi}}{\sqrt{d_i^2 + d_j^2 - 2d_i d_j \cos(\theta_{ij})}}\right) - \sum_{n=1}^{\infty}\frac{\left(\frac{d_i d_j}{R_{Pi}}\right)^2 \cos(n\theta_{ij})}{n}\right)$$

(3.100)

The existence of potential coefficients between the pipe and the ground depends on the pipe thickness. If the pipe thickness is considered infinite, all the entries are zero and the other potential coefficients matrices are reduced as the last row and column are eliminated. If the pipe thickness is not considered infinite and part of the current returns in the ground, the matrix is calculated according to (3.101).

$$[P_c] = \begin{bmatrix} P_{\text{pipe}} & P_{\text{pipe}} & P_{\text{pipe}} & P_{\text{pipe}} & P_{\text{pipe}} & P_{\text{pipe}} & P_{\text{pipe}} \\ P_{\text{pipe}} & P_{\text{pipe}} & P_{\text{pipe}} & P_{\text{pipe}} & P_{\text{pipe}} & P_{\text{pipe}} & P_{\text{pipe}} \\ P_{\text{pipe}} & P_{\text{pipe}} & P_{\text{pipe}} & P_{\text{pipe}} & P_{\text{pipe}} & P_{\text{pipe}} & P_{\text{pipe}} \\ P_{\text{pipe}} & P_{\text{pipe}} & P_{\text{pipe}} & P_{\text{pipe}} & P_{\text{pipe}} & P_{\text{pipe}} & P_{\text{pipe}} \\ P_{\text{pipe}} & P_{\text{pipe}} & P_{\text{pipe}} & P_{\text{pipe}} & P_{\text{pipe}} & P_{\text{pipe}} & P_{\text{pipe}} \\ P_{\text{pipe}} & P_{\text{pipe}} & P_{\text{pipe}} & P_{\text{pipe}} & P_{\text{pipe}} & P_{\text{pipe}} & P_{\text{pipe}} \\ P_{\text{pipe}} & P_{\text{pipe}} & P_{\text{pipe}} & P_{\text{pipe}} & P_{\text{pipe}} & P_{\text{pipe}} & P_{\text{pipe}} \end{bmatrix}$$

(3.101)

$$P_{\text{pipe}} = \frac{1}{2\pi\varepsilon_p} \ln\left(\frac{R_{\text{Pins}}}{R_{Po}}\right)$$

(3.102)

3.3.3.1 Example

A Matlab code where the several parameters are calculated is available at the book's webpage.

The cable used in the example is equal to the one used in example 3.3.1.1, but surrounded by a pipe, whose parameters are shown in Table 3.2. The layout of the cable is identical of the one shown in Fig. 3.7.

All the parameters used in the calculation of the series impedance matrix in the previous example are valid, with the exception of the Z_{earth}. The new parameters are calculated below.

$$ZSPinsul = 0 + j3.187 \times 10^{-5}\,\Omega/m$$
$$ZPinner = 2.219 \times 10^{-5} + j8.221 \times 10^{-6}\,\Omega/m$$
$$ZPouter = 1.850 \times 10^{-5} + j2.267 \times 10^{-6}\,\Omega/m$$
$$ZPGinsul = 0 + j9.686 \times 10^{-6}\,\Omega/m$$
$$ZPmutual = 1.839 \times 10^{-5} + j1.196 \times 10^{-6}\,\Omega/m$$
$$ZPipe\_mutual = 1.683 \times 10^{-5} + j1.655 \times 10^{-5}\,\Omega/m$$
$$Zearth = 4.945 \times 10^{-5} + j5.530 \times 10^{-4}\,\Omega/m$$

The series impedance is given by (3.103).

$$[Z] = \begin{bmatrix} 0.080+j0.668 & 0.053+j0.609 & 0.048+j0.584 & 0.048+j0.584 & 0.048+j0.584 & 0.048+j0.584 & 0.050+j0.566 \\ 0.053+j0.609 & 0.232+j0.608 & 0.048+j0.584 & 0.048+j0.584 & 0.048+j0.584 & 0.048+j0.584 & 0.050+j0.566 \\ 0.048+j0.584 & 0.048+j0.584 & 0.080+j0.668 & 0.053+j0.609 & 0.048+j0.584 & 0.048+j0.584 & 0.050+j0.566 \\ 0.048+j0.584 & 0.048+j0.584 & 0.053+j0.609 & 0.232+j0.608 & 0.048+j0.584 & 0.048+j0.584 & 0.050+j0.566 \\ 0.048+j0.584 & 0.048+j0.584 & 0.048+j0.584 & 0.048+j0.584 & 0.080+j0.668 & 0.053+j0.609 & 0.050+j0.566 \\ 0.048+j0.584 & 0.048+j0.584 & 0.048+j0.584 & 0.048+j0.584 & 0.053+j0.609 & 0.232+j0.608 & 0.050+j0.566 \\ 0.050+j0.566 & 0.050+j0.566 & 0.050+j0.566 & 0.050+j0.566 & 0.050+j0.566 & 0.050+j0.566 & 0.068+j0.565 \end{bmatrix}\ m\Omega/m$$

$$(3.103)$$

The admittance matrix is calculated using the potential coefficients matrices (3.97)–(3.102). The calculated terms are given below and the matrix in (3.104).

$$P_{cond} = 4.127 \times 10^{9}\,\Omega$$

$$P_{screen} = 8.216 \times 10^{8}\,\Omega$$

$$P_{Pii} = 3.039 \times 10^{9}\,\Omega$$

$$P_{Pij} = 2.571 \times 10^{9}\,\Omega$$

$$P_{pipe} = 9.236 \times 10^{8}\,\Omega$$

Table 3.2 Pipe-type cable data

Layer	Radius (mm)
R_{Pi}	107
R_{Po}	120
R_{Pins}	140

$\rho_{pipe} = 1.71 \times 10^{-8}\ \Omega m$; $\mu_{pipe\_rel} = 1$; $\varepsilon_{pipe\_rel} = 3$

$$[Y] = \begin{bmatrix} j0.0761 & -j0.0761 & 0 & 0 & 0 & 0 & 0 \\ -j0.0761 & j0.1825 & 0 & -j0.0308 & 0 & -j0.0308 & -j0.0449 \\ 0 & 0 & j0.0761 & -j0.0761 & 0 & 0 & 0 \\ 0 & -j0.0308 & -j0.0761 & j0.1825 & 0 & -j0.0308 & -j0.0449 \\ 0 & 0 & 0 & 0 & j0.0761 & -j0.0761 & 0 \\ 0 & -j0.0308 & 0 & -j0.0308 & -j0.0761 & j0.1825 & -j0.0449 \\ 0 & -j0.0449 & 0 & -j0.0449 & 0 & -j0.0449 & j0.4747 \end{bmatrix} \mu S/m$$

$$(3.104)$$

If the pipe thickness is infinite, the series impedance matrix and admittance matrix are given by respectively (3.105) and (3.106).

$$[Z] = \begin{bmatrix} 0.049 + j0.100 & 0.022 + j0.042 & 0.017 + j0.017 & 0.017 + j0.017 & 0.017 + j0.017 & 0.017 + j0.017 \\ 0.022 + j0.042 & 0.201 + j0.041 & 0.017 + j0.017 & 0.017 + j0.017 & 0.017 + j0.017 & 0.017 + j0.017 \\ 0.017 + j0.017 & 0.017 + j0.017 & 0.049 + j0.100 & 0.022 + j0.042 & 0.017 + j0.017 & 0.017 + j0.017 \\ 0.017 + j0.017 & 0.017 + j0.017 & 0.022 + j0.042 & 0.201 + j0.041 & 0.017 + j0.017 & 0.017 + j0.017 \\ 0.017 + j0.017 & 0.017 + j0.017 & 0.017 + j0.017 & 0.017 + j0.017 & 0.049 + j0.100 & 0.022 + j0.042 \\ 0.017 + j0.017 & 0.017 + j0.017 & 0.017 + j0.017 & 0.017 + j0.017 & 0.022 + j0.042 & 0.201 + j0.041 \end{bmatrix} m\Omega/m$$

$$(3.105)$$

$$[Y] = \begin{bmatrix} j0.076 & -j0.076 & 0 & 0 & 0 & 0 \\ -j0.076 & j0.183 & 0 & 0 & 0 & 0 \\ 0 & 0 & j0.076 & -j0.076 & 0 & 0 \\ 0 & 0 & -j0.076 & j0.183 & 0 & 0 \\ 0 & 0 & 0 & 0 & j0.076 & -j0.076 \\ 0 & 0 & 0 & 0 & -j0.076 & j0.183 \end{bmatrix} \mu S/m$$

$$(3.106)$$

3.3.4 Summary

This section demonstrated and explained the methods normally used to calculate the *series impedance* and *admittance* matrices of a cable bonded in both ends or cross-bonded.

The calculation of the series impedance matrix is made using loop equations that correspond to loops between the conductors of each phase (core, screen and armour). A transformation matrix is then used to convert the *loop impedance matrix* to the correspondent *series impedance matrix* typically used when analysing a system.

The *admittance matrix* is easier to calculate as the electric field is normally confined to each phase and there is no electric coupling between phases.

It was also demonstrated how to calculate the matrices for pipe-type cables, the differences between considering the thickness of the pipe finite or infinite and that a thickness can normally be considered as infinite, as the penetration depth is smaller than the typical thickness of a pipe.

Two examples show how to calculate the matrices for a three-phase single-core cable and a pipe-type cable.

3.4 Modal Analysis

It is now clear that the analysis of cables is more complex than the analysis of OHLs, as we have normally to work with systems that have six to nine coupled equations. Thus, it would be very helpful if we could decouple these equations which could then be solved as single-line equations.

This is achieved by changing the phase equations to the modal domain as explained next. This method was originally designed for OHLs and later adapted to underground cables. The mathematical processes are identical, but the process is usually more complicated for cables because of the higher number of propagation modes.

3.4.1 Method

We start by adapting the wave Eqs. (3.14) and (3.15), previously studied in 3.2, to respectively (3.107) and (3.108).[3]

$$\left[\frac{d^2 V_{ph}}{dx^2}\right] = [Z_{ph}][Y_{ph}][V_{ph}] \tag{3.107}$$

$$\left[\frac{d^2 I_{ph}}{dx^2}\right] = [Y_{ph}][Z_{ph}][I_{ph}] \tag{3.108}$$

The objective is to change the system from a coupled system to a decoupled system, i.e. from the phase domain to the modal domain, something that is made using the eigenvalues and eigenvectors theory. The relation between the voltage in the phase and modal domains is given by (3.109), where T_V is the transformation matrix. Using the relation (3.109), (3.107) can be written as (3.110).

$$[V_{ph}] = [T_V][V_M] \tag{3.109}$$

$$\left[\frac{d^2 V_M}{dx^2}\right] = [T_V]^{-1}[Z_{ph}][Y_{ph}][T_V][V_M] \tag{3.110}$$

---

[3] As we are dealing with matrices, the order of the matrices cannot be arbitrary.

The transformation matrix T_V diagonalises the product $[Z_{ph}]$ $[Y_{ph}]$, and it is therefore the eigenvectors of this product, whereas the eigenvalues are equal to the diagonal elements of (3.111). As a result, (3.110) can be written as (3.112).

$$[\Lambda] = [T_V]^{-1} [Z_{ph}] [Y_{ph}] [T_V] \tag{3.111}$$

$$\left[\frac{d^2 V_M}{dx^2} \right] = [\Lambda][V_M] \tag{3.112}$$

A similar reasoning can be applied to the current and it is concluded that the transformation matrix is related to the transformation matrix of the voltage by (3.113).

$$T_I = [T_V]^{-T} \tag{3.113}$$

The series impedance matrix $[Z_{ph}]$ and the admittance matrix $[Y_{ph}]$ are both frequency dependent and complex numbers. Thus, the transformations matrices, $[T_V]$ and $[T_I]$, and the eigenvalues are also frequency dependent and complex numbers.

A complex matrix is decoupled if and only if it is *normal*, i.e., M*M = MM*, where M* is the *conjugate transpose* of M. In other words, the matrix has to be either *Hermitian* or *unitary*.

The product of $[Z_{ph}]$ and $[Y_{ph}]$ matrix can never be Hermitian, as these matrices must have real terms in the main diagonal.

Also, the product of the matrices cannot be unitary as it would be necessary for the columns vectors as for the rows vectors to form an orthonormal basis, something which is also in principle not possible, except for some notable cases.

Thus, it seems that the modal decomposition is not that efficient, as the modes continue to be coupled, and the transformation matrices are therefore not constant and vary with the frequency.

This is mathematically true, but as the frequency increases, the imaginary part of the product of the matrices come closer to zero (see Fig. 3.11), and the matrix can be considered as close to real and in the form (3.114). In these conditions, a three-phase single-core cable can be diagonalised.

Fig. 3.11 Angle of the eigenvalues in function of the frequency

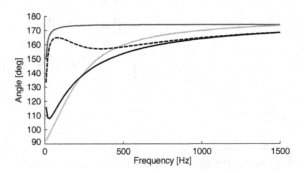

$$\begin{bmatrix} a & 0 & 0 & b & e & e \\ 0 & a & 0 & e & b & e \\ 0 & 0 & a & e & e & b \\ c & 0 & 0 & d & e & e \\ 0 & c & 0 & e & d & e \\ 0 & 0 & c & e & e & d \end{bmatrix} \quad (3.114)$$

Thus, with some simplifications, we can consider the system decoupled for frequencies above a few kHz at the cost of a small inaccuracy, typically less than 10% for frequencies above 200 Hz and 1% for frequencies over 1 kHz.

3.4.1.1 Example

A Matlab code where are calculated the transformation matrices, the speed and the attenuation for both SC cable and pipe-type is available at the book's webpage.

As an example, we are going to calculate the transformation matrices of the three-phase single-core cable described in example 3.3.1.1.

The series impedance matrix and admittance matrix are calculated as in the previous example. In order to keep the congruency with other authors and sources, the entries of both matrices change position as shown in (3.115). The cores of the three phases are now aggregated in the first three columns/rows and the screens are next.

$$Z_{Ph} = \begin{array}{cccccc} Core\,1 & Core\,2 & Core\,3 & Screen\,1 & Screen\,2 & Screen\,3 \\ \begin{bmatrix} Z_{C1C1} & Z_{C1C2} & Z_{C1C3} & Z_{C1S1} & Z_{C1S2} & Z_{C1S3} \\ Z_{C2C1} & Z_{C2C2} & Z_{C2C3} & Z_{C2S1} & Z_{C2S2} & Z_{C2S3} \\ Z_{C3C1} & Z_{C3C2} & Z_{C3C3} & Z_{C3S1} & Z_{C3S2} & Z_{C3S3} \\ Z_{S1C1} & Z_{S1C2} & Z_{S1C3} & Z_{S1S1} & Z_{S1S2} & Z_{S1S3} \\ Z_{S2C1} & Z_{S2C2} & Z_{S2C3} & Z_{S2S1} & Z_{S2S2} & Z_{S2S3} \\ Z_{S3C1} & Z_{S3C2} & Z_{S3C3} & Z_{S3S1} & Z_{S3S2} & Z_{S3S3} \end{bmatrix} \end{array} \quad (3.115)$$

Considering a frequency high enough for the system to be decoupled, the voltage transformation matrix and the current transformation matrix (3.116) are obtained.[4]

---

[4] Numerically calculated for 20 kHz.

$$[T_V] \simeq \begin{bmatrix} 1/\sqrt{6} & 0 & 1/\sqrt{3} & 1/\sqrt{3} & 2/\sqrt{6} & 0 \\ 1/\sqrt{6} & 1/2 & -1/(2\sqrt{3}) & 1/\sqrt{3} & -1/\sqrt{6} & 1/\sqrt{2} \\ 1/\sqrt{6} & -1/2 & -1/(2\sqrt{3}) & 1/\sqrt{3} & -1/\sqrt{6} & -1/\sqrt{2} \\ 1/\sqrt{6} & 0 & 1/\sqrt{3} & 0 & 0 & 0 \\ 1/\sqrt{6} & 1/2 & -1/(2\sqrt{3}) & 0 & 0 & 0 \\ 1/\sqrt{6} & -1/2 & -1/(2\sqrt{3}) & 0 & 0 & 0 \end{bmatrix}$$

$$[T_I] \simeq \begin{bmatrix} 0 & 0 & 0 & 1/\sqrt{3} & 2/\sqrt{6} & 0 \\ 0 & 0 & 0 & 1/\sqrt{3} & -1/\sqrt{6} & 1/\sqrt{2} \\ 0 & 0 & 0 & 1/\sqrt{3} & -1/\sqrt{6} & -1/\sqrt{2} \\ 2/\sqrt{6} & 0 & 2/\sqrt{3} & -1/(\sqrt{3}) & -2/\sqrt{6} & 0 \\ 2/\sqrt{6} & 1 & -1/(\sqrt{3}) & -1/(\sqrt{3}) & 1/\sqrt{6} & -1/\sqrt{2} \\ 2/\sqrt{6} & -1 & -1/(\sqrt{3}) & -1/(\sqrt{3}) & 1/\sqrt{6} & 1/\sqrt{2} \end{bmatrix}$$

$$(3.116)$$

As previously explained, the transformation matrices are frequency dependent and have a high variation at low frequencies. Figure 3.12 gives an example by showing the magnitude of the elements of column 5 of $[T_I]$ in function of the frequency, where it is observed how the entries are very frequency dependent for the low frequencies region.

At this point, it is important to physically analyse the results and to describe the six propagation modes of (3.116). The propagation modes of a three-phase single-core armourless cable are divided into:

- 3 coaxial modes between the conductor and the sheath of each phase (Fig. 3.13).
- 2 intersheath modes between the sheaths (Fig. 3.14).
- 1 ground mode between the sheaths and the ground (Fig. 3.15).

The coaxial modes correspond to the core-screen loops, in which the current in the core fully returns in the screen of the same cable.

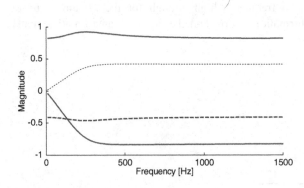

Fig. 3.12 Magnitude of the elements of columns 5 of $[T_I]$

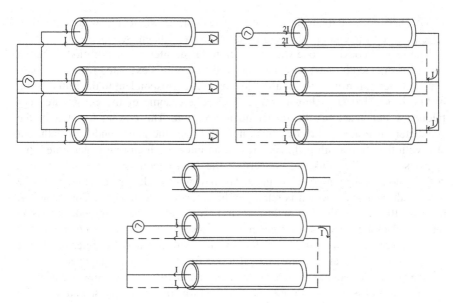

Fig. 3.13 Coaxial modes

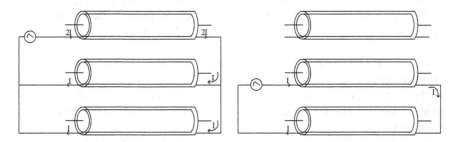

Fig. 3.14 Intersheath modes

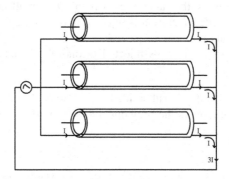

Fig. 3.15 Ground mode

The first coaxial mode is a zero-sequence mode where the same current is injected in all three cores and returns in the correspondent screens.

The second coaxial mode was expected to be an inter-conductor mode where a current is injected in one phase and half of the current is extracted from one phase and the other half from other phase. This mode is correspondent to the fifth column of $[T_I]$ in (3.116) also shown in Fig. 3.12, where some of the entries are very frequency dependent in the low-frequency region. The current injected in the conductor generates a flux that links not only with the inter-conductor path, but also with the intersheath path. As a result, an intersheath current is generated that opposes to the flux linkage, converting the inter-conductor mode into both a coaxial mode and an intersheath mode. The induced circulating current depends on the sheath resistance, which is relatively high for low frequencies. Consequently, there is little current flowing in the screen for low frequencies and mode acts as an inter-conductor mode. As the frequency increases, the relative impedance of the screen decreases[5], the current in the screens increases and the inter-conductor mode becomes virtually a pure coaxial mode for the higher frequencies.

The third coaxial mode is also an inter-conductor mode at low frequencies where a current is injected in one phase and returns in another phase and a coaxial mode at high frequencies. The behaviour of this mode is similar to the second coaxial mode and only at a high frequency does it behave as a pure coaxial mode.

The intersheath modes correspond to the screen-screen loops where the current flowing in one screen returns in one or both screens of the other cables.

The first intersheath mode is obtained by injecting a current in the screen of one phase and extracting half of it from one phase and the other half from the other phase.

The second intersheath mode is obtained by injecting a current in the screen of one phase and extracting it from another phase.

The ground mode corresponds to the screen-ground loop, and it is a zero-sequence mode where the same current is injected in all three screens and returns in the ground.

By reading the description of the modes, it is seen that another possible division of them would be to consider:

- 2 zero-sequence modes: the same current is injected in all three phases;
- 2 inter-conductor modes: a current is injected in one phase and subdivides itself by the other two phases;
- 2 conductor modes: a current is injected in one phase and returns in another phase.

Each of the three modes is subdivided into a coaxial and sheath mode. This division is not as common as the previous, but it is used by some authors.

---

[5] This does not mean that the screen impedance decreases with the frequency. It means that when compared with the other impedances it is relatively lower for high frequencies.

Pipe-Type Cable

We have seen that the analysis of pipe-type cables is subdivided into two: finite and infinite thickness.

If the thickness is infinite, the analysis is mathematically equal to the one previously done, but the ground mode changes to pipe mode.

If the thickness is finite, the series impedance and admittance matrices increase from 6×6 to 7×7, leading to increase in the number of modes from six to seven, the extra being mode the pipe mode, which now exists together with the ground mode.

Cable with armour

If each of the phases has an armour, the modes increase from six to nine, sub-divided as follows:

- 3 coaxial (core) modes,
- 3 coaxial (sheath) modes,
- 2 interarmour modes,
- 1 ground mode.

The logic is identical to the one previously explained, but with three new modes that represent the sheath-armour loop of each phase and a change from intersheath mode to interarmour mode.

However, there is a special case that requires our attention. Quite often the armour of the cable may be made of steel or other materials with a high permeability.

In these cases, the magnetic field is more confined to the interior of the cable, affecting the mutual coupling between phases and the modal matrices.

The current transformation matrix (3.117) is for a cable with an armour, whose relative permeability is 1, whereas the current transformation matrix (3.118) is for a relative permeability of the armour equal to 400. Both transformation matrices are obtained for 20 kHz.

In the first case, the matrix is similar to the ones shown for an armourless cable. The first column shows the ground mode, the second and third columns the interarmour mode, the fourth to seventh column the coaxial modes of the core and the remaining columns the coaxial modes of the sheath.

In the second case, the results are rather different and the similarities are found only for the ground mode and the interarmour modes. We can see that for the remaining modes the current inject into the conductor of a phase returns all in the screen of the same phase, whereas the current inject into the screen of a phase returns all in the armour of the same phase.

$$[T_I] \simeq \begin{bmatrix} 0 & 0 & 0 & 1/\sqrt{3} & 2/\sqrt{6} & 0 & 0 & 0 & 0 \\ 0 & 0 & 0 & 1/\sqrt{3} & -1/\sqrt{6} & 1/\sqrt{2} & 0 & 0 & 0 \\ 0 & 0 & 0 & 1/\sqrt{3} & -1/\sqrt{6} & -1/\sqrt{2} & 0 & 0 & 0 \\ 0 & 0 & 0 & -1/\sqrt{3} & -2/\sqrt{6} & 0 & 2/\sqrt{6} & 2/\sqrt{3} & 0 \\ 0 & 0 & 0 & -1/\sqrt{3} & -/\sqrt{6} & -1/\sqrt{2} & 2/\sqrt{6} & -1/\sqrt{3} & 1 \\ 0 & 0 & 0 & -1/\sqrt{3} & 1/\sqrt{6} & 1/\sqrt{2} & 2/\sqrt{6} & -1/\sqrt{3} & -1 \\ 1 & 0 & \sqrt{2} & 0 & 0 & 0 & -2/\sqrt{6} & -2/\sqrt{3} & 0 \\ 1 & -\sqrt{6}/2 & -1/\sqrt{2} & 0 & 0 & 0 & -2/\sqrt{6} & 1/\sqrt{3} & -1 \\ 1 & \sqrt{6}/2 & -1/\sqrt{2} & 0 & 0 & 0 & -2/\sqrt{6} & 1/\sqrt{3} & 1 \end{bmatrix}$$

$$(3.117)$$

$$[T_I] \simeq \begin{bmatrix} 0 & 0 & 0 & 1 & 0 & 0 & 0 & 0 & 0 \\ 0 & 0 & 0 & 0 & 1 & 0 & 0 & 0 & 0 \\ 0 & 0 & 0 & 0 & 0 & 1 & 0 & 0 & 0 \\ 0 & 0 & 0 & -1 & 0 & 0 & \sqrt{2} & 0 & 0 \\ 0 & 0 & 0 & 0 & -1 & 0 & 0 & \sqrt{2} & 0 \\ 0 & 0 & 0 & 0 & 0 & -1 & 0 & 0 & \sqrt{2} \\ 1 & 0 & \sqrt{2} & 0 & 0 & 0 & -\sqrt{2} & 0 & 0 \\ 1 & \sqrt{6}/2 & -\sqrt{2}/2 & 0 & 0 & 0 & 0 & -\sqrt{2} & 0 \\ 1 & -\sqrt{6}/2 & -\sqrt{2}/2 & 0 & 0 & 0 & 0 & 0 & -\sqrt{2} \end{bmatrix} \quad (3.118)$$

3.4.2 Modal Velocities

We have seen that the modes propagate in different materials (core, screen, armour and ground). As a result, the modes have different speeds. This difference is very important because during a transient, all the modes are excited and the associated waveform depends in part on the velocities of each propagation mode.

The velocity of each mode is calculated using the eigenvalues of the product of $[Z_{Ph}]$ by $[Y_{Ph}]$ (3.119).

$$v = \frac{2\pi f}{\mathrm{imag}\left(\sqrt{(\Lambda)}\right)} \tag{3.119}$$

The coaxial modes correspond to the current flowing in the core-screen loop and for high frequencies they have a velocity that is approximately equal to the propagation velocity of the cable, normally between 160 and 180 m/μs. The velocity is lower at the lower frequencies, because the inductance is larger for those frequencies.

The intersheath modes have a lower velocity than the coaxial modes, typically between 40 and 80 m/μs. The speed of the intersheath mode is very dependent on mutual earth impedance, which is very often calculated using approximations. Thus, a small variation in the impedance can represent a not negligible difference in the velocities of these modes.

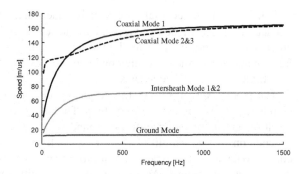

Fig. 3.16 Velocity of the modes of propagation in function of the frequency

The relation between the coaxial mode and intersheath mode velocities is given by (3.120) (see [8] for the mathematical development). The intersheath velocity depends on the distance between phases, and it decreases when the distance increases. As a result, the intersheath velocity of a cable in flat formation is lower than if the cable was installed in a trefoil formation. The velocities of the inter-sheath modes 1 and 2 are also distinct in the latter.

$$V_{IS} \simeq \sqrt{\frac{\ln\left(\frac{R_4}{R_3}\right)}{\ln\left(\frac{d}{R_3}\right)}} V_C \qquad (3.120)$$

The ground mode has the lower velocity, typically between 10 and 20 m/µs. The lower velocity is because of the high reactance of the ground return path, which can be 100 times larger than that of the cable.

Figure 3.16 shows the velocity of the several modes in function of the frequency for the cable used in the example of the previous section.

Pipe-Type Cable

The mathematical procedure is in all identical to the one used to single-core cables, but the results are substantially different.

Figure 3.17 shows the velocity of the different modes. Comparing with the single-core case, it is seen that the ground mode was substituted by a pipe mode with a lower speed, that the velocity of the intersheath modes reduces substantially and that the coaxial modes are barely affected, except for the lower frequencies.

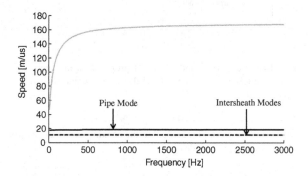

Fig. 3.17 Velocity of the modes of propagation in function of the frequency for a pipe-type cable with infinite pipe thickness

Starting with the coaxial modes, it is understandable to have only small differences as these modes correspond to the core-screen loops, which we have previously seen are not affected by the pipe or the ground. The differences at low frequencies are in the two interconductor modes that behave partially as intersheath modes at low frequencies.

The pipe mode substitutes the ground mode, meaning that the screen-ground loop is replaced by the screen-pipe loop. The low velocity of the ground mode is a result of the high reactance of the ground return path. The reactance of the pipe depends on its permeability, which is typically very large. As an example, it is normal to have a relative permeability for the pipe of 200–400, whereas the one of the core and screen is 1. As a result, for this specific example, the velocity of the pipe mode is even lower than that of the ground mode in the previous example.

The decrease in the speed of the intersheath modes is explained by the changes in the impedance of the return path, which affects the screen-screen loops (see Sect. 3.3). In other words, the reason behind the decrease of the velocity in the intersheath modes is the same behind the lower velocity of the pipe mode when compared with the ground mode.

It is also important to notice that as the permeability of the pipe reduces the velocity increases. If the relative permeability reduced to 1, the velocity of the pipe and intersheath modes would be approximately the same of the coaxial modes.

Cable with armour

There are nine different modes for a cable with armour, in opposition to the six of an armourless cable. Figure 3.18 shows the modal velocities for an armour with a relative permeability of 1, and an armour with a relative permeability of 400.

We can observe the ground and interarmour modes that have low velocities, as expected, for both permeabilities. We can also observe that while the velocity of the coaxial core modes is approximately 160 m/μs even for lower frequencies, the velocity of coaxial sheath modes only achieves those values for the very high frequencies.[6]

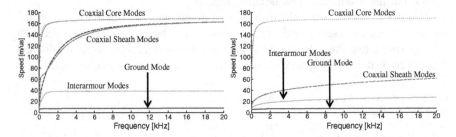

Fig. 3.18 Velocity of the modes of propagation in function of the frequency for a cable with armour. **a** Relative permeability of the armour equal to 1. **b** Relative permeability of the armour equal to 400

---

[6] It is not shown in Fig. 3.18b, but the coaxial sheath modes would eventually reach the velocity of the coaxial core modes.

3.4.3 Modal Attenuation

In the same way that the modes have different velocities, they also have different attenuations, meaning that the damping is not the same for all the modes.

The attenuation of each mode is calculated using the real part of the eigenvalues (3.121). Figure 3.19 shows the attenuation of each mode in function of the frequency, using a logarithmical scale.

$$\alpha = \text{real}\left(\sqrt{(\Lambda)}\right) \tag{3.121}$$

The analysis of the results shows that the attenuation of each mode increases with the frequency, which for the higher frequencies is explained by skin effect. At the lower frequencies, the modes are not completely decoupled and have a mutual influence as we have previously seen.

An individual analysis of each mode shows that the ground mode has a larger attenuation than the other modes.The reason for this behaviour is the high resistivity of the ground, which can be thousands of million times larger than the resistivities of the conductor and/or screen.

The high attenuation of the intersheath modes at low frequencies is because the ground is the return path of this mode at the lower frequencies.

The coaxial modes have the lower attenuation, because the resistance of the conductor is also the lower when compared with the screen and ground.

Pipe-Type Cable
Figure 3.20 shows the attenuation of the different modes of the pipe-type cable with a pipe of infinite thickness used in the previous examples.In this case the attenuation of the coaxial modes and the pipe mode are similar, being the one of the intersheath modes slightly lower. The substitution of the ground mode by the pipe mode would have to mean a large reduction of the attenuation of the latter when compared with the former, because of the much lower resistivity. As all three resistivities, conductor, screen and pipe, are similar, the attenuations also tend to be more alike.

The permeability of the pipe also influences the attenuation, even if not much. The skin depth decreases when the permeability increases, thus, the larger the

Fig. 3.19 Attenuation of the modes of propagation in function of the frequency

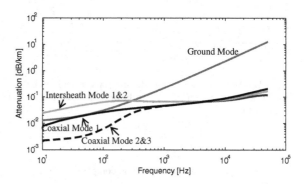

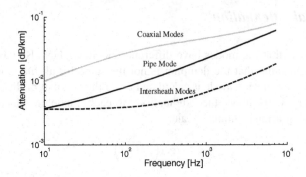

Fig. 3.20 Attenuation of the modes of propagation in function of the frequency for a pipe-type cable with infinite pipe thickness

permeability the lower the attenuation of the pipe mode. At lower frequencies there will also be more variations in function of the permeability because of the nature of the interconductor and intersheath modes, which is not so visible for a high pipe permeability.

Cable with armour
Figure 3.21 shows the attenuation of the different propagation modes in function of the frequency. As expected the larger attenuation is for the ground mode. An interesting result is the larger attenuation of the coaxial screens modes and the interarmour modes when compared with the armourless cable example.[7] This increase is explained by the larger resistivity of the steel in the armour, which results also in a larger resistance.

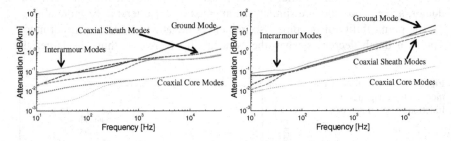

Fig. 3.21 Attenuation of the modes of propagation in function of the frequency for a cable with armour. **a** Relative permeability of the armour equal to 1. **b** Relative permeability of the armour equal to 400

---

[7] In this case the interarmour mode is compared with the intersheath mode.

3.4.4 Summary

In this section, we have seen how to calculate the modal quantities of an underground cable. The importance of these modes will become clearer later in the book when we analyse transient waveforms and their simulation.

We have also seen how the modes are frequency dependent and that the velocity and attenuation of each mode depends on the frequency. Unfortunately, the transformation matrices are also frequency dependent, and only at high frequencies can one say that the modes are decoupled and the transformation matrices are constant.

The number of modes depends on the number of conductors in the system, but it is typically six (nine if each phase of the cable has an armour), which are subdivided into three coaxial modes, two intersheath/interarmour modes and one ground/pipe mode.

3.5 Frequency Spectrum of a Cable

We know intuitively that a cable has a lower resonance frequency than an equivalent OHL, because of its much larger capacitance. However, the bonding configuration can also influence the frequency spectrum.

The shunt admittance matrix is equal for both bonding configurations as the distance between the conductor and the screen is unchanged, but the series impedance matrix presents some differences.

In applying the method explained in Sect. 3.3 (3.122) and (3.123) are obtained for the calculation of positive-sequence series impedance of the conductor for both-ends bonding and cross-bonding, respectively.

$$Z_{\text{both}}^{+} = Z_{CC} - Z_{CS2} + \frac{(Z_{CS1} - Z_{CS2})^2}{Z_{CS2} - Z_{SS}} \tag{3.122}$$

$$Z_{\text{cross}}^{+} = Z_{CC} - Z_{CS2} \tag{3.123}$$

We will see that the resonance points have higher magnitudes and lower frequencies for the cross-bonded cable than for the both-end bonded cable. As the differences in the resonances are noticeable for the first resonance point, the nominal pi-model instead of the equivalent pi-model can be used ,[8] which results in a simple mathematical analysis.

The line impedance for the nominal pi-model is given by (3.124).

---

[8] The differences between the two models are typically not very large up to the first resonance point, becoming more noticeable as the frequency increases.

$$Z = \frac{(1 - \omega^2 LC) + j(\omega RL)}{(-\omega^2 C^2 R) + j(2\omega C - \omega^3 C^2 L)} \tag{3.124}$$

The shunt admittance is equal for both bonding types and the differences in the impedance are a function of L and R.

Resonance Frequency

The resonance frequency is given by (3.125), which is obtained by developing (3.124). L is the imaginary part of the series impedance (3.122)–(3.123). Consequently, the imaginary part of $Z^+_{cross} > Z^+_{both}$. Subsequently, the imaginary part of the last element of (3.122) should be negative, i.e. (3.126).

$$\omega^2 = \frac{2}{LC} \tag{3.125}$$

$$\text{imag}\left(\frac{(Z_{CS1} - Z_{CS2})^2}{Z_{CS2} - Z_{SS}}\right) < 0 \tag{3.126}$$

Both the real and imaginary parts of Z_{SS} are always larger than the equivalents in Z_{CS2}. Thus, the denominator of (3.126) has always a negative real and imaginary part (3.127).[9]

$$Z_{CS2} - Z_{SS} = -a - jb \tag{3.127}$$

The development of the numerator of (3.126) results in (3.128).

$$\begin{aligned}(Z_{CS1} - Z_{CS2})^2 &= ((c + jd) - (e + jf))^2 \\ &= (c^2 - d^2 + e^2 - f^2 - 2ce + 2df) + j(2cd + 2ef - 2cf - 2de)\end{aligned} \tag{3.128}$$

The magnetic field is stronger between the core and screen of the same cable than between the cores or core-screen of two different cables. Thus, $d > f$ while $c \approx e$, and so (3.128) can be simplified to (3.129).

$$\begin{aligned}(Z_{CS1} - Z_{CS2})^2 &= (-d^2 - f^2 + 2df) + j0 \Leftrightarrow (Z_{CS1} - Z_{CS2})^2 = \left(-d^2 - (d - g)^2 + 2d(d - g)\right) \\ &\Leftrightarrow (Z_{CS1} - Z_{CS2})^2 = -d^2 - d^2 - g^2 + 2dg + 2d^2 - 2dg \Leftrightarrow (Z_{CS1} - Z_{CS2})^2 = -g^2\end{aligned} \tag{3.129}$$

It is concluded that the numerator of (3.126) has only a negative real part. Therefore, (3.126) can be written as (3.130).

---

[9] The variables a to g are real numbers that are used in the mathematical demonstration.

$$\frac{(Z_{CS1} - Z_{CS2})^2}{Z_{CS2} - Z_{SS}} = \frac{-g^2}{-a - jb} = -g^2 \cdot (-a' + jb')$$

$$\Leftrightarrow \frac{(Z_{CS1} - Z_{CS2})^2}{Z_{CS2} - Z_{SS}} = (a'g^2) + j(-b'g^2)$$

(3.130)

The imaginary part of (3.126) is always negative and the resonance frequency of a cross-bonded cable is always lower than the resonance frequency if bonded at both ends.

Magnitude at the parallel resonance frequency
From (3.124) the magnitude is given by (3.131).

$$\|Z\| = \frac{(1 - \omega^2 LC) + j\omega RC}{-\omega^2 RC^2} \Leftrightarrow \|Z\| = \frac{(1 - \omega^2 LC)}{-\omega^2 RC^2} + \frac{j\omega RC}{-\omega^2 RC^2}$$

$$\Leftrightarrow \|Z\| = \frac{L}{2RC} + j\frac{L}{C^2\sqrt{\frac{2}{LC}}} \Leftrightarrow \|Z\| = \sqrt{\frac{L^2}{4C^2R^2} + \frac{L^3}{2C^3}}$$

(3.131)

$$\Leftrightarrow \|Z\| = \sqrt{\left(\frac{L}{C}\right)^2 \frac{1}{4R^2} + \left(\frac{L}{C}\right)^3 \frac{1}{2}}$$

The two variables that depend on the bonding are R and L, which are, respectively, the real and imaginary parts of the series impedance matrix (to be precise the imaginary part is XL, but for this analysis, that is not very relevant).

From the analysis of the resonance frequency together with (3.122), (3.123) and (3.130), it is known that the both-end bonded cable has a higher resistance and a lower inductance. Doing the substitutions in (3.131), it is concluded that the magnitude of the parallel resonance points is lower in the both-end bonded cable.

Magnitude at the series resonance frequency
For a series resonance the impedance magnitude is given by (3.132). The value of L is lower for a both-end bonded cable, resulting in lower magnitude at the series resonance points for this type of bonding.

$$Z = \frac{j(\omega RL)}{-\omega^2 C^2 R} \Leftrightarrow \|Z\| = \frac{L}{\omega C^2} \Leftrightarrow \|Z\| = \frac{L\sqrt{LC}}{2C^2}$$

(3.132)

In summary, the conductor positive-sequence series inductance is larger in a cross-bond cable than in an equivalent cable bonded at both ends, whereas the series resistance is larger for a cable bonded at both ends.

These differences in the series impedance result in lower resonance frequencies for a cross-bond cable. They are also responsible for the larger impedance magnitude at the parallel resonance points and lower impedance magnitude at the series resonance points of the cross-bond cable, when compared to the cable bonded at both ends.

From a physical point of view, the larger inductance in the cross-bond cable is a result of having a lower current circulating in the screen when compared to the

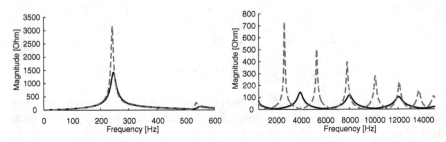

Fig. 3.22 Frequency spectrums seen from the LEM node with the cable open in the STSV end. *Dashed line* cross-bonded with 12 major sections, *solid line* both-ends bonding

cable at both ends. As the current is lower, the magnetic field induced by screen current is also lower, resulting in a larger inductance value.

We have demonstrated how the bonding configurations affect the frequency spectrum of cable, it is now time to see some examples and if software simulations show the expect results

Figure 3.22 shows the impedance spectrum considering a 20 km cable as being bonded at both ends or cross-bonded with 12 major sections. The cable is connected in series to a transformer in order to have a resonance point at a low frequency with a large magnitude, resultant of the cable-transformer interaction, so that we can verify if the bonding configuration also affects the resonance between a cable and other equipment.

The resonance frequency seen around 250 Hz is the cable-transformer resonance. The resonance points seen in the second figure for frequencies higher than 600 Hz are the cable resonance frequencies.

The first resonance point is at approximately 250 Hz or the 5th harmonic. At this frequency, there is a large difference between the magnitude of the impedances, whereas the resonance frequency is almost the same for both configurations.

As the frequency increases, the differences between the two bonding configurations start to become more evident:

- The cross-bonded cable has more resonance points than the cable bonded at both ends;
- The magnitude of the cross-bonded cable impedance is larger at the parallel resonance points and lower at the series resonance points, when compared with the cable bonded at both ends;

Both results confirm the theory.

Comparison of cross-bonded cables with different number of sections
The number of major sections of a cross-bonded cable can vary. The theoretical deduction previously done assumes an ideal cross-bonding, i.e. a perfect balancing that it only achieved with an infinite number of sections, something that is impossible. Thus, our next step is to study how the existence of a limit number of major sections may affect the frequency spectrum.

Fig. 3.23 Comparison of the frequency spectrum in PSCAD/EMTDC. *Solid line* one major cross-bonded sections, *dashed line* six major cross-bonded sections, *dotted line* 12 major cross-bonded sections

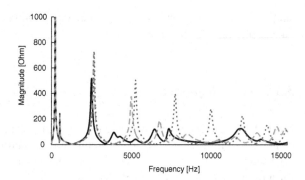

Figure 3.23 compares the frequency spectrum for a different number of major sections, for the system previously shown in Fig. 3.22.

The higher the number of cross-bonded sections, the closer the results are to the ideal cross-bonding. Using the 12 major cross-bonded sections scenario as reference, it is seen that the cable with only one major section starts to diverge after the first parallel resonance point (~ 2.5 kHz) and the one with six major sections after the third parallel resonance point (~ 7 kHz).

After these frequencies, the respective spectrums present an unexpected behaviour. They have more resonance points, whose magnitude does not always decrease with the increase in frequency.

This behaviour is the result of the larger imbalance present when less cable sections are used. Figure 3.24 shows the voltage in the cable receiving end when injecting a 134.5 kV peak voltage in the sending end at different frequencies. The figure shows the results for a cable bonded in both ends and an equivalent cable with a major cross-bonded section.

When observing the cross-bonded cable spectrum, it can be seen that between two larger peak voltages (upper dashed circles) two smaller overvoltages (lower dashed circles) are present, which are the result of the two crossing of the screens.

The higher the number of major cable cross-bonded sections, the more balanced the cable and the coupling are. As a result, the entire cable behaves like a uniform

Fig. 3.24 Voltage in the cable receiving end for a 134.5 kV peak voltage in the cable sending end. *Solid line* bonded in both ends, *dashed line* one major cross-bonded section

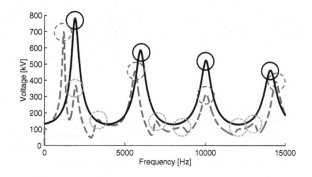

single section; if only one major cross-bonded section is present, the cable behaves almost like three different cables.

3.5.1 Zero-Sequence

Most of the zero-sequence components of a buried cable return through the screen of the cable, influencing the short-circuit currents and the TRVs. As a result, the zero-sequence current in the screen is roughly equal to the current in the conductor for any random given point of the cable.

Consequently, there should not be almost no current flowing into the ground at the grounding points[10] and the frequency spectrums should be independent of the type of bonding and number of major sections.

Figure 3.25 shows the frequency spectrums for several bonding configurations.

The resonance frequencies are the same for all the examples because of the reasons explained in the previous two paragraphs. The both-end bonded cable and the cable with only one major section have the same magnitude at all times, but the same it not the case for a cable with more major sections that have a different magnitude for some of the resonance frequencies.

The reactance of an open line, either cable or OHL, is not the same at all points of the line. The reactance can be described by a hyperbolic function and it will have a theoretical infinite value to lengths that are multiple of half of the wavelength.[11]

A cable with multiple cross-bonded sections also has multiple grounding points. If one of those grounding points corresponds to a point of the line where the

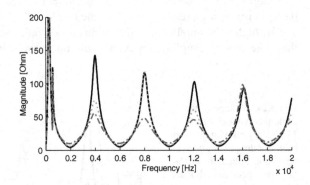

Fig. 3.25 Frequency spectrum for the D2-eq system. *Dashed line* both-ends bondings, *solid line* one major cross-bonded section, *dotted line* six major cross-bonded sections, *dashed-dotted line* 12 major cross-bonded sections

---

[10] Short circuits are special cases as we will see in Sect. 4.10.

[11] The opposite case happens for a quarter wavelength, which corresponds to very low impedances. If the line is short circuited in the end the opposite occurs, very high impedance at a quarter wavelength and very low at half wavelength.

(a)   **(b)**

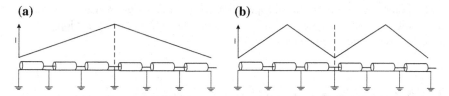

Fig. 3.26 Relative current along a cable with six major cross-bonded sections. **a** 1st parallel resonance point (~ 4 kHz). **b** 2nd parallel resonance point (~ 8 kHz)

reactance has a very high value, the current in the screen flows to the ground at that point, changing the magnitude of the impedance.

Example

The example is given for the cable with six major sections. For a parallel resonance situation, the maximum current in the cable occurs for the point(s) that are equal to or a multiple of the cable wavelength, i.e. $\lambda/2$; $3\lambda/2$; $5\lambda/2$ ....

Figure 3.26 shows on the top part the relative current for the first two parallel resonance frequencies along a cable drawn on the lower part of the figure. For the first resonance frequency, the maximum current is in the middle of the cable line. For the second resonance frequency, the peak currents are in the first and last quarter of the cable line.

In the first case, the screen is grounded in a point corresponding to maximum current, which means a very high reactance in that point. Thus the current flows to the ground in that point, but this does not happen in the second case, which have the peak currents in the middle point between two grounds.

As a consequence of this, and using the both-end bonded cable as reference, the magnitude of the impedance is smaller for the first resonance point (~ 4 kHz) and equal for the second resonance point (~ 8 kHz).

It should be noted that the presence of other conductors near the cable, as an earth continuity conductor, can change the results presented in this section as these conductors present a possible path for the zero-sequence components.

3.5.2 Summary

The type of bonding used affects the positive-sequence resonance frequencies, whereas the influence in the zero-sequence spectrums is only on the magnitude of the resonance points. A cross-bonded cable has more resonance frequencies than an equivalent cable bonded at both ends.

However, the type of bonding has little influence in the event of a cable-transformer resonance, which occurs for frequencies lower than the cable's first resonance frequency.

For other phenomena, like propagation of harmonics generated by different harmonic sources, the type of bonding used is relevant, especially for long cables with low resonance frequencies.

3.6 Reflection and Refraction of Travelling Waves

An electromagnetic wave propagating along a line, either cable or OHL, is associated to a correspondent voltage/current waves pair. The relation between the voltage and current waves is constant and it is commonly known as *characteristic impedance* or *surge impedance* (Z_0).

The characteristic impedance of a line depends on its parameters and frequency (3.133). Thus, each type of line has different characteristic impedances, as the parameters depend on the materials used in the construction of the line, the geometry and the layout.

$$Z_0(\omega) = \frac{\sqrt{R + j\omega L}}{\sqrt{G + j\omega C}} \tag{3.133}$$

When an electromagnetic wave arrives to a discontinuity point, e.g. a transition from a cable to an OHL or a line termination, there must an adjustment of the voltage and the current as there are changes in the characteristic impedance.

Like in any physical system, there must be conservation of energy. Therefore, when a wave reaches a discontinuity point, part of the energy of the wave propagates past the discontinuity, whereas the remaining energy is reflected back into the line.[12] In a similar way, the refracted voltage must be equal to the injected voltage plus the reflected voltage and the same for the current.

The refracted voltage in a discontinuity point is given by (3.134), where V_1 is the injected voltage, V_2 is the reflected voltage, V_3 the refracted voltage and Z_A and Z_B the surge impedance of the lines. The reflected voltage is calculated using (3.135). The currents are calculated by dividing the voltages by the respective characteristic impedances or by using (3.136) and (3.137).

$$V_3 = V_1 \frac{2Z_B}{Z_A + Z_B} \tag{3.134}$$

$$V_2 = V_1 \frac{Z_B - Z_A}{Z_A + Z_B} \tag{3.135}$$

$$I_3 = I_1 \frac{2Z_A}{Z_A + Z_B} \tag{3.136}$$

---

[12] Assuming the system is lossless.

$$I_2 = I_1 \frac{Z_A - Z_B}{Z_A + Z_B} \tag{3.137}$$

Example

In order to understand the phenomenon, it is consider a cable-OHL line system. A DC electromagnetic step wave of 1 pu is injected into the cable, and it decomposes into two waves when the discontinuity point is reached, i.e., the cable-OHL junction.

Typically, a cable characteristic impedance is approximately 50 Ω, whereas in an OHL, it is 400 Ω. Thus, the refracted voltage is equal to 1.778 pu (3.138) and the reflected voltage to 0.778 pu (3.139). The refracted and reflected currents are 0.222 and −0.778 pu, respectively .

$$V_3 = V_1 \frac{2Z_B}{Z_A + Z_B} \Leftrightarrow V_3 = 1 \frac{2 \cdot 400}{50 + 400} \Leftrightarrow V_3 = 1.778 \, \text{pu} \tag{3.138}$$

$$V_2 = V_1 \frac{Z_B - Z_A}{Z_A + Z_B} \Leftrightarrow V_2 = 1 \frac{400 - 50}{400 + 50} \Leftrightarrow V_2 = 0.778 \, \text{pu} \tag{3.139}$$

In this example, the voltage increases when flowing from the cable into the OHL, whereas the current decreases. The opposite would happen if the wave was flowing from the OHL into the cable.

This result is very important as some systems have hybrid cable-OHL lines where the voltage and current peak values during the transient are strongly related to the discontinuity and the difference in the characteristics impedances (more on this on Sect. 4.8).

One common mistake made by many students is to consider that the voltage in the first line after the reflection is only the reflected voltage, whereas in reality is the summation of the injected and reflected waves, by other words the reflected wave is superimposed to the injected to wave. Another way of seeing the phenomenon is that it is necessary to have continuity of the current and voltage, meaning that the voltage and current have to be equal on both sides of the discontinuity.

Figure 3.27 shows the voltage and the current before and after the wave reached the discontinuity point for a DC wave.

For AC waves, the analysis is more complicated as the waves change over time. For power frequency and typical lengths, the travelling time is so high that wave magnitude can be considered constant during the duration of the entire phenomenon. However, for high frequencies, it is necessary to consider the changes over time of the voltage and current when superimposing both waves.

A discontinuity point can also be a convergence point of three or more lines, in which the injected wave is refracted into two or more lines.

The reasoning applied in this scenario is similar to the ones previously explained. There must be continuity of the voltage, meaning that all the refracted

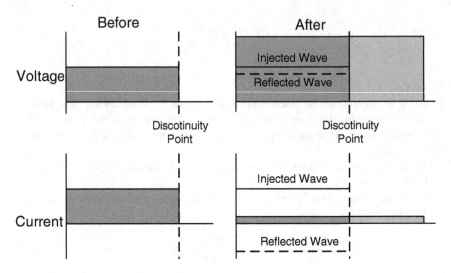

Fig. 3.27 Propagation of a DC wave in a discontinuity point

voltages and injected voltage plus the reflected voltage have to be equal (3.140). The continuity of the current results in the summation of all the refracted currents to be equal to the injected current plus the reflected current (3.141).

$$V_1 + V_2 = V_{3a} = V_{3b} = V_{3c} \cdots \tag{3.140}$$

$$I_1 + I_2 = I_{3a} + I_{3b} + I_{3c} + \cdots \tag{3.141}$$

The explanations given in this chapter are according to the classic theory that fits better with OHLs. A detailed analysis and study of the propagation of electromagnetic waves in HVAC is more complex, mainly because of the current that flows in the screen of the cable. Section 4.3 will provide a detailed explanation on reflections in HVAC cables and on how the bonding configuration affects the waves, together with several examples.

3.6.1 Line Terminations

A line termination is also a discontinuity point and Eqs. (3.134)–(3.137) can continue to be used to calculate the reflection coefficient. However, if the termination is inductive or capacitive, it is necessary to write Z_B in the Laplace domain and solve the equation. If we think about it, it makes sense, as what we are doing is charging an inductor or capacitor, similar to what we have seen in Chap. 2. Thus, the refracted wave is equal to the cases studied in Chap. 2 and the reflected wave is the refracted wave minus the injected wave.

Two of the more notable cases are a line terminating in an open end or a short circuit. If the line terminates in an open end, the characteristic impedance of Z_B is infinite. By applying (3.135) and (3.137), it is concluded that for an open end, the reflected voltage is equal to the injected voltage and the reflected current has the same magnitude of the injected current, but opposite polarity.

If the line terminates in a short circuit, the characteristic impedance of Z_B is zero and the behaviour is the opposite of an open termination. The reflected current is equal to the injected current and the voltage switches polarity while maintaining the magnitude.

A generator can also be considered a line termination. If the generator is ideal, i.e. infinite short-circuit power, the voltage wave is reflected back into the line with the same magnitude and opposite polarity while the current is reflected with the same magnitude and polarity.

3.7 Summary

This chapter provided the basic tools that we will use to simulate and understand the electromagnetic transient phenomena analysed in the next chapters. We started by studying the classic Telegraphs equations that explain the travelling waves in a conductor.

These equations can be adapted to study the waves travelling in a HVAC cable, but it is first necessary to write the *series impedance* and *admittance* matrices of the cable. The matrices are obtained using the loop theory that is thoroughly explained in the chapter.

The mathematical and physical analyses of the cable would be easier if the equations were decoupled. Such operation is possible and demonstrated in the chapter.

Finally, the chapter recapitulated the basis of wave reflections paving the way for the deeper study made in the next chapter.

3.8 Exercises

1. Consider that the cable of example 3.3.1.1 is cross-bonded and has three minor sections, each with a length equal to one-third of the cable used in the example. Calculate the series impedance matrix for each minor section and the series impedance for the entire cable, considering ideal cross-bonding, i.e. like if there was an infinite number of sections, so that there would be perfect balance.
2. Calculated the loop matrix for the cable of example 3.3.1.1 and apply the transformation matrix to it. Compare with the matrix presented in the example.

3. Repeat the exercise for an armour cable and calculate also the admittance matrix.

Layer	Thickness (mm)
Armour (galvanised steel)	5
Insulation over the screen	2

$\rho_{Armour} = 9.1 \times 10^{-7} \; \Omega m^{-1}$; $\mu_{Armour\_rel} = 1{,}000$; $\varepsilon_{Armour\_rel} = 1$

4. The matrices shown in example 3.4.1.1 consider the system has being fully decoupled. Calculate the matrices for 50 and 200 Hz.
5. A 10 km cable is connected to a 6 km OHL. The electrical parameters of the cable and OHL are:

   Cable: R = 0 Ω/km; L = 0.4 mH/km; C = 0.16 μF/km

   OHL: R = 0 Ω/km; L = 1.6 mH/km; C = 10 nF/km

   Consider the system as single phase and energised at peak voltage (1 pu) at t = 0 s. Design the lattice diagram of the voltage during the first 190 μs.

References and Further Reading

1. Popović Z, Popović BD (2000) Introductory electromagnetics. Prentice Hall, New Jersey
2. Greenwood A (1991) Electrical transients in power systems, 2nd edn. Wiley, New York
3. Marshall SV, Dubroff RE, Skitek GG (1996) Electromagnetic concepts and applications, 4th edn. Prentice Hall, New Jersey
4. Tleis N (2008) Power systems modelling and fault analysis: theory and practice. Elsevier, Oxford
5. Martinez-Velasco JA (2010) Power system transients: parameter determination. CRC Press, Boca Raton
6. van der Sluis L (2001) Transients in power systems. Wiley, New York
7. Dommel HW (1986) Electro-magnetic transients program (EMTP) theory book. Bonneville Power Administration, Portland
8. Wedepohl LM, Wilcox DJ (1973) Transient analysis of underground power-transmission systems: system-model and wave-propagation characteristics. In: Proceedings of the institution of electrical engineers, vol 120(2)
9. Brown GW, Rocamora RG (1976) Surge propagation in three-phase pipe-type cables, Part I: unsaturated pipe. IEEE Trans. Power Apparatus Syst PAS-95(1)
10. Ametani A (1980) Wave propagation characteristics of cables. IEEE Trans. Power Apparatus Syst PAS-99(2)
11. Ametani A (1980) A general formulation of impedance and admittance of cables. IEEE Trans. Power Apparatus Syst PAS-99(3)
12. Noualy JP, Le Roy G (1977) Wave-propagation modes on high-voltage cables. IEEE Trans. Power Apparatus Syst PAS-96(1)
13. Nagaoka N, Ametani A (1983) Transient calculations on crossbonded cables. IEEE Trans. Power Apparatus Syst PAS-102(4)

14. Morched A, Gustavsen B, Tartibi M (1999) A universal model for accurate calculation of electromagnetic transients on overhead lines and underground cables. IEEE Trans Power Delivery 14(3)
15. Yang Y, Ma J, Dawalibi FP (2001) Computation of cable parameters for pipe-type cables with arbitrary thicknesses. In: IEEE-PES transmission and distribution conference and exposition
16. Noda T (2008) Numerical techniques for accurate evaluation of overhead line and underground cable constants. IEEJ Trans Electr Electron Eng
17. Gudmundsdóttir US (2010) Modelling of long high voltage AC cables in transmission systems. PhD Thesis, Aalborg University
18. CIGRE WG B1.30 (2012) Cable systems electrical characteristics. CIGRE, Paris

Chapter 4
Transient Phenomena

4.1 Introduction

Several topics, from Laplace equations to modal theory, were explained in the previous chapters building moment for this chapter where several electromagnetic transient phenomena are finally described.

We will start by describing switching overvoltage in great detail, laying bases that we can use in the analysis of other phenomena and situations.

After that we will study several phenomena from the more typical to some unusual cases that are specific for cable-based networks. In the website are available PSCAD simulations for each phenomenon that can help understanding the different phenomena.

4.2 Different Types of Overvoltages

The IEC standards [1, 2] define overvoltages into four different types:

- Temporary overvoltages (TOV). TOV are characterised by their amplitude, voltage shape and duration, as they may have a long duration, up to one minute. These overvoltages can be caused by faults, switching conditions, resonance conditions and non-linearities.
- Slow-Front Overvoltages (SFO). SFO are characterised by their voltage shape and amplitude, they have a front duration of up to some milliseconds and are oscillatory by nature. SFO can be caused by line energisations and de-energisations, faults, switching of capacitive/inductive elements (e.g., cables and shunt reactors) or distant lightning strokes.
- Fast-Front Overvoltages (FFO). FFO are mainly associated with lightning strokes and are characterised by the standard lightning impulse wave (1.2/50 μs).
- Very-Fast-Front Overvoltages (VFFO). VFFO are associated with GIS switching operations and SF_6 circuit breaker re-ignitions, topics that are not dealt with in this book.

F. F. da Silva and C. L. Bak, *Electromagnetic Transients in Power Cables*,
Power Systems, DOI: 10.1007/978-1-4471-5236-1_4,
© Springer-Verlag London 2013

The phenomena described in this book fall into the three first overvoltage types, especially TOV and SFO.

4.3 Switching Overvoltage

The analysis of transient phenomena starts with the simplest of all, the switching overvoltage. We will deeply study this phenomenon as it will lay the bases for the following phenomena.

To give a first a first idea of the phenomenon, we start by considering the cable as an LC load connected to an ideal voltage source. In this system, both capacitor and inductor are initially discharged, but as the voltage in the capacitor (Vc) has to be continuous when the circuit breaker closes for a voltage value that is not zero, the capacitor has to be charged through the inductor, initiating a transient. After a short moment, the voltage in the capacitor is equal to the source voltage, but at that instant the current in the inductor is at peak value (see Fig. 4.1) and by energy conservation it cannot become zero immediately. Thus, the voltage in the capacitor continues to increase, exceeding the source voltage, while the current decreases to zero, when the current crosses zero, Vc reaches a peak value and the capacitor starts to discharge.

The frequency of the voltage in the capacitor and in the inductor is not power frequency, but the system resonant frequency. As a result, the voltage in the capacitor (Vc) matches the source voltage at different points. Thus, the magnitude of Vc is different for each resonant cycle, as the reference voltage for the capacitor terminals is constantly changing.

The initial increase of the voltage in the capacitor is proportional to the difference between the voltage in the source and the voltage in the capacitor at the

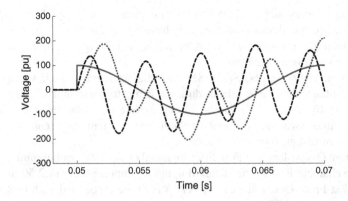

Fig. 4.1 Voltages and current (current not at scale) for a LC load when connected at peak voltage. *Solid line* voltage in the source, *dotted line* voltage in the capacitor, *dashed line* current in the load

energisation instant. If the CB is closed at zero voltage there is no voltage difference between the capacitor and the source at the energisation instant and consequently there is no transient. In similar way, if the capacitor is not discharged and the voltage at the switching instant matches the "voltage" in the capacitor,[1] there is also no transient.

4.3.1 Single-Core Cables

The somehow simplistic explanation given in the previous section helps to have a first idea of the phenomenon, but it is far from giving a realistic image. To do that it is necessary to use travelling waves theory.

We start by considering a 50 km long one phase single-core cable energised by an ideal voltage source at peak voltage. The cable is energised at peak voltage, meaning that a 1 pu voltage impulse is injected into the cable. After approximately 0.3 ms,[2] the voltages reach the receiving end of the cable and it is fully reflected back. In a lossless line, the voltage at the receiving end would be 2 pu at this instant, but because of the damping the value is substantially smaller. Consequently, it can also be concluded that the shorter the cable is the less the damping and the larger the first peak voltage.

After another 0.3 ms, the wave reaches the cable sending end and it is reflected back into the cable. In a real system, the voltage at the sending end of the cable would be distorted at this instant (see Sect. 4.3.3), but we are using an ideal voltage source in this example and consequently the voltage in the sending end is imposed by the source. However, we can see when the wave arrives to the sending end by observing the current sudden drop in that end. In a lossless system, the magnitude of the current after the reflection would be equal to the magnitude before the reflection with opposite polarity. In reality, the magnitude is lower because of the losses.

After another 0.3 ms, i.e., 3×0.3 ms after the energisation, the wave that was reflected back into the cable at the sending end reaches the cable receiving end, i.e. the original wave impulse reaches the receiving end of the cable for a second time. The voltage wave was reflected back with a negative polarity at the ideal voltage source installed in the sending end. Consequently, the voltage at the receiving end is now reduced.

The reflections continue until the transient be completely damped by the cable resistance. Figure 4.2 shows the voltage and the current in a cable during its energisation at peak voltage. The waveforms described in the previous paragraphs can be observed in the figure.

---

[1] Given by the $V = Q/C$ relation.

[2] Given by approximately 50,000 m/150 m/μs.

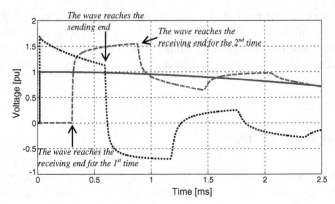

Fig. 4.2 Voltages and current (current not at scale) when energising a cable at peak voltage. *Solid line* voltage in sending end, *dashed line* voltage in the receiving end, *dotted line* current in the sending end

The figure shows that the voltage in the receiving end does not jump immediately to its final value having instead a continuous increasing until the initial wave reaches the receiving end for the second time. The current in the sending end has a similar behaviour decreasing continuously until the transient wave reaches the sending end after be reflected in the receiving end. This behaviour is a result of the inductance and capacitance of the line and it was explained for lumped parameters models in Chap. 2.

As previously stated, the transient behaviour does not depend directly on the magnitude of the source voltage, but on the difference between the voltage in the cable and source at the energisation instant. Normally, a cable is discharged before being energised and the worst case scenario, i.e., largest transient, is for an energisation at peak voltage whereas the best case scenario is an energisation at zero voltage, when there is virtually no transient.

Yet, if the cable is not discharged, the energisation at zero voltage may no longer be the best case scenario. A cable is mainly capacitive and there is almost a 90° phase difference between the current and the voltage. A CB opens its contacts when the current is crossing zero and consequently the voltage has approximately maximum magnitude at the disconnection instant, the exception are CBs prepare to interrupt currents with several amperes, but this type of CB is not common. For sake of simplicity, we assume that the cable is not discharged after the switch-off and that the voltage remains constant after the disconnection.[3]

Thus, if the cable is energised at a peak voltage of polarity equal to the one of the cable, there is almost no difference between the voltage in the cable and the source at the energisation instant and there is barely a transient. Figure 4.3a shows an example of such energisation where it can be seen that there is virtually no transient, the small transient shown in the zoom-box is consequence of having a

---

[3] In reality the cable would discharge, but that takes several seconds or even minutes.

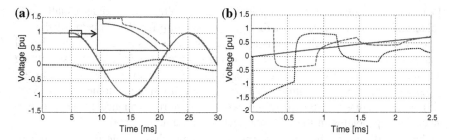

Fig. 4.3 Voltages and current (current not at scale) when energising a charged cable. *Solid line* voltage in sending end, *dashed line* voltage in the receiving end, *dotted line* current in the sending end. **a** Energisation at peak voltage. **b** Energisation at zero voltage

small increase in the voltage at the cable's sending end immediately after the disconnection,[4] ideally, it would not be present. On the other hand to energise this cable at zero voltage would result in a transient in all similar to the one previously described, as seen in Fig. 4.3b.[5]

By now some of readers have noticed that if the energisation was made for peak voltage and opposite polarity, the transient would be even worse, as the initial voltage difference would have been 2 pu. This is true, but we will leave that case for the study of transient recovery voltages in Sect. 4.7.

4.3.2 Three-Phase Cables

The cases studied in the previous section were for a single-phase cable. The energisation of the more usual three-phase cables is similar to the energisation of a single-phase cable and the same principles and theories can be applied in its study. Yet, there are some minor differences and details that must also be addressed.

Two types of switching are commonly used when energising a three-phase cable. The first and older is the classic switching using a ganged CB where all three phases are connected at the same instant. In this situation, the phases have different transient waveforms as the voltage is not same in all three at the energisation instant. Moreover, it is normally not possible to control the switching instant.

The second type of switching is synchronised switching, also known as single-pole switching, and it consists in energising the phases at different instants.

---

[4] The voltage is not the same all along the cable at the disconnection instant, being typically lower at the sending end of the cable, when assuming that the other end is already disconnected. After the disconnection, the voltage in the cable tends to equalise resulting in a voltage increase at the sending end of the cable (for more information on the phenomenon see Sect. 4.10.1).

[5] In this case, the voltage difference is ~ -1 pu at the energisation instant, and the voltage variations have opposite polarity to the ones shown in Fig. 4.2.

Typically, each phase is energised for the same voltage magnitude, usually zero volts, and the waveforms in all three phases are similar. The control of the energisation instant for each phase is a big advantage as it is possible to energise all phases at zero volts and avoid the switching overvoltages in all three, something that it is not possible when using a ganged breaker. However, other problems are raised by the use of synchronise switching, as zero-missing phenomenon (see Sect. 4.5) or an increase in the likelihood of having ferroresonance because of the mechanical problems in the CB (see Sect. 4.9.3), among others.

Another difference between energising a single-phase cable or a three-phase cable is in the mutual coupling between phases, which has a small influence in the peak magnitudes.

We start by thinking in a case where all three-phases are energised at the same instant, with phase A energised at peak voltage. The voltage in the other two phases at the energisation instant is equal to −0.5 pu (± 120° phase difference).

The transient current generated in phases B and C have an opposite polarity to the current of Phase A, meaning that current, and consequently the voltage in phase A, increase because of mutual coupling with these two phases, as given by Lenz's Law. Figure 4.4a shows an example by comparing the voltage in the receiving end and the current in the sending end for Phase A of a three-phase cable and an equivalent single-phase cable both energised at peak voltage. The maximum peak voltage in the three-phase cable is 0.102 pu larger than in the single-phase cable, a 6.6 % increase. Different configurations will give different results, but a voltage increase can usually be expected.

The mutual coupling between phases influences also the energisations made by means of synchronise switching. Not only because of the effect described in the previous paragraph, but also because the voltage in the cables at the energisation instant is no longer precisely 0 V for the phases that are energised after the first. As a result, the voltage difference between the source and the cable at the energisation instant may be slightly larger than expected, resulting in a larger peak voltage.

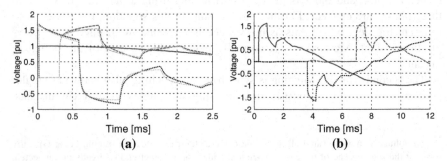

Fig. 4.4 a Voltages and current (current not at scale) when energising a cable at peak voltage. *Solid line* voltage in sending end, *dashed line* voltages in the receiving end, *dotted line* currents in the sending end. **b** Energising of a cable using synchronise switching with all phases energised at peak voltage

Figure 4.4b shows an example of such case, where the magnitude of the peak voltages in phases A, B and C are, respectively, 1.611, 1.648 and 1.65 pu.

4.3.3 Modelling of the Source

The modelling of the source and the value of the short-circuit power at the sending end of the cable influence the waveform shape, the magnitude of the peak voltage and the duration of the overvoltage.

In this section, we analyse how the reflections in the sending end are affected by the modelling of the source, but considering only strong networks, as the ones typically found in Europe.

In the previous examples, we consider the voltage source as being ideal and with an infinite short-circuit power. We will now simulate the energisation of the same cable considering a 10,000 MVA short-circuit power. We subdivide the energisation in two subcases: a first where an equivalent RL network is used and a second case where the cable is integrated in the model of a real transmission network with dozens of buses and lines.

We start with the RL equivalent network, which consist basically in having a lumped resistance and inductance between the voltage source and the cable. In this situation, the cable is energised through the inductance and the current has to be continuous as seen in Chap. 2. Consequently, the magnitude of the voltage wave flowing into the cable is not immediately equal to the magnitude of the source voltage; instead, the voltage rises to the source value with a time constant that is roughly given by L/Z_0, where Z_0 is the characteristic impedance of the cable. As a result, the waveform no longer provides an accurate image of reality as the peak voltage and current are both wrong[6] and the waveform behaves like a mix of lumped parameters and travelling waves showing an exaggerated resonance frequency.

The second subcase consists in energising the cable using the model of a transmission network as feeding network. By other words, there is not a voltage source, but several generators, which are distant from the cable. Between the generators and the cable, there are several cables/OHLs, transformers, shunt reactors and other equipment that is typical found in a real network. The network's voltage level is 165 kV and the short-circuit power at the cable sending end is approximately 10,000 MVA.

In this situation, the waveforms are much closer to the ideal source case, but with some differences. The first and main difference is that the impulse sent into the cable is smaller than the voltage at the sending end at the switching instant.

---

[6] They are typically 10–15% larger than in reality.

Fig. 4.5 Voltage in the receiving end of the cable for different equivalent networks. *Solid line* ideal voltage source, *dotted line* RL equivalent network, *dashed line* transmission system model

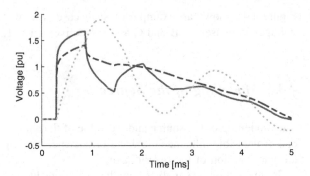

When a cable is connected to other cables, the initial instantaneous voltage at the switching instant divides itself by all the cables. As an example, if an unenergised cable, only connected to an already energised cable with an identical surge impedance, is energised for an initial voltage of 1 pu, a 0.5 pu voltage wave propagates into the cable being energised and a −0.5 pu wave propagates into the energised cable. In this situation, the peak voltage has a theoretical maximum of 1 pu, half of the value that would be obtained using an ideal voltage source.[7] If the surge impedances were different, the voltage would divide according to the respective values.

In this specific subcase, the switched cable is connected to two cables with different surges impedance and a transformer, and the voltage at the switching instant is 1.1 pu. At the closing of the CB, a wave of approximately 0.78 pu is injected into the cable and a wave of approximately −0.32 pu into the adjacent cables; the voltage in the sending end reduces from 1.1 to 0.78 pu. The 0.78 pu wave is doubled at the receiving end of the cable and the theoretical maximum peak voltage is 1.56 pu, in reality is 1.41 pu because of damping.

The second difference is the reflection of the wave at the sending end of the cable. In the ideal voltage source example, the wave is reflected back into the cable with a reflection coefficient of −1 pu, but in this case the reflection coefficient is smaller, in absolute value, and dependent on the surge impedance of the cable and adjacent cables, as explained in Sect. 3.6. Consequently, the voltage drop at the receiving end of the cable occurring when of the original wave reaches the receiving end for the second time is much smaller than for the ideal voltage source case.

Figure 4.5 shows the voltage in the receiving end for an energisation with an ideal voltage source, a RL equivalent network and a model of the transmission system. The figure shows a visual example of the phenomena explained in the previous pages. Using the ideal voltage source case as reference, the wave correspondent to the RL equivalent network has a larger peak voltage and resonance behaviour, whereas the case using the model of a transmission network has a lower

---

[7] This theoretical maximum does not consider the reflections in other adjacent cables, i.e., the switched cable is much shorter than the cable connected to its sending end. The more normal situation of similar lengths is explained in Sect. 5.2.

peak voltage, due to the smaller injected wave, and lower voltage gradient, because of the lower reflection coefficient at the sending end.

4.3.4 Influence of the Bonding

We have previously studied the energisation of a cable, but with focus only in the conductor. Yet, the bonding configuration has a strong influence in the waveforms and the peak voltage and current. However, for several reasons, the modelling of all cross-bonded sections represents a substantial increase in both the simulation running time and the time necessary to design the system:

- It is necessary to design n minor cross-bonded sections instead of 1/3;
- The software needs more time to complete each time step;
- The time step has to be reduced in order to accommodate the shorter cable sections;
- Increases the probability of fitting and memory problems;

Thus, we would prefer to model as little sections as possible. The existing IEC standard (IEC TR-60071-4) suggests the modelling of all cross-bonded sections. Some phenomena require this level of detail, whereas for other phenomena the model can be simplified. Yet, before we learn when and how the model can be simplified we have to understand how the bonding affects the transients, mainly the switching transient.

The explanation is given below, but with some simplifications, because of the high complexity of the phenomenon. More specifically, it is not consider mutual coupling between phases and the values are rounded.

For simplicity, we start by analysing a cable bonded in both-ends. Figure 4.6 shows the voltages and currents at different points of the conductor and screen, for a cable energised at peak voltage by means of synchronise switching. The waves are shown for the sending end, the end of the first section (1/3 of the cable) and the end of the second section (2/3 of the cable), i.e., where there would be the cross between minor-sections in a cross-bonded cable. The explanation of the waveforms is given after the figures.

The analysis of the waves is made using the modal theory previously described in Sect. 3.4. We start with the analysis of the current waveforms (top figures in Fig. 4.6).

The current in the screen is first seen at the end of the first section (point A in Fig. 4.6), at precisely the same instant that the current in the conductor reaches that point. The modal analysis tells us that for the coaxial mode, the current in the screen is equal to the current in conductor, but with opposite direction. Comparing the two currents, we see that they have virtually the same magnitude and opposite polarities, confirming the theory. Slightly later, the wave arrives to the end of the second section of the cable (point B) and the same behaviour is observed.

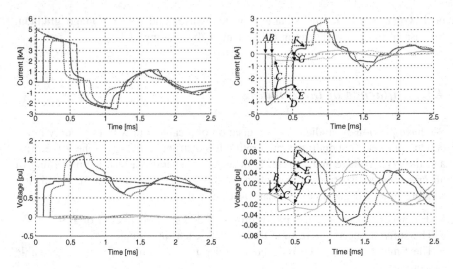

Fig. 4.6 Currents (*top*) and voltages (*bottom*) in a cable bonded in both-ends. *Left* conductor, *right* screen. *Dashed line* sending end, *solid line* end of the first section, *dotted line* end of the second section

Until this instant the current in the screen matches the behaviour of the current in the conductor. Then, at point C, there is a sudden variation of the current in the screen at the first section that is not seen in the conductor's current. This variation is explained by the arrival of the intersheath mode generated at the energisation instant, more precisely the second intersheath mode (the interconductor mode) where the current flowing in the screen subdivides itself and returns in the other two screens. Figure 4.6 shows that the current in screen of the phase being energised increases approximately 700 A, whereas the current in the other two screens decreases approximately 350 A, confirming the expected.

We calculated in previous examples the speed of the coaxial modes (~ 170 m/μs) and the intersheath mode (~ 70 m/μs) of the cable used in this example. Each of the sections used in the example is 16.666 km long. Thus, it is expected that the coaxial mode wave reaches the end of the first section after ~ 0.1 ms and the intersheath mode after ~ 0.240 ms; similar values are obtained in the simulation.

Points D and E show a reduction of the current that is a result of the current in the conductor be reflected back at the end of the cable with opposite polarity. As a result, the current in the conductors reduces to almost zero and the one in the screens becomes positive because it adds to the intersheath mode current that is still propagating to the cable receiving end. If there was no intersheath mode current, the current in the screens would reduce to a value close to zero, being slightly negative in order to have the symmetry with the currents in the conductors, resultant of the coaxial mode.

Point G shows the arrival of the intersheath mode current of point C to the end of the second section.

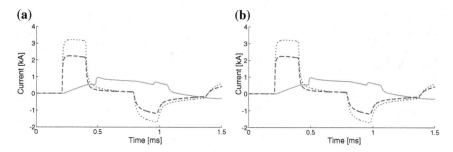

Fig. 4.7 Modal currents during the cable energisation. **a** End of the first section. **b** End of the second section. *Dashed line* coaxial mode 1, *dotted line* coaxial mode 2, *solid line* intersheath mode 1

Point *F* shows the current reflected in the sending and reaching the end of the first section as explained in Sects. 4.3.1 and 4.3.2.

Figure 4.7 confirms the results and the explanations just given by showing the modal currents during the energisation. There are two coaxial modes, the ones correspond to Phase 1, and one intersheath mode, whereas the remaining modes are zero. The variation of the modal current corresponds to the variation seen in the phase currents.

The voltage waveforms are also explained using the modal theory. The coaxial modes do not affect the voltage in the screen (Eq. (3.116) in Sect. 3.4). Thus, there is only a small variation of the voltage in the screens at point *A*, in opposition to what happens with the current. Ideally, there would not be any variation, but for that the model components would have to be fully decoupled.

However, the intersheath mode changes the voltage in the conductor and it is seen a variation at Point *B*. As for the current, the increase of the voltage in the screen of the phase being energised is double and opposite in polarity to the gradient of the other two phases. An important difference is that the intersheath mode affects also the voltage in the conductor, something that does not happen for the current.[8] The analysis of the matrix T_V shows that the voltage variation in the screen and conductor should be the same and the simulation confirms the theory.

The remaining waveform behaviour can now be easily deduced similar to what was done for the current.

We stop now the analysis of the cable bonded in both-ends and we start analysing a cross-bonded cable with equal length. Similar to what was done for the cable bonded in both-ends, the cross-bonded cable is divided into three sections of equal lengths. The screens of the cable are transposed between the minor-section, resulting in a cross-bonded cable with one major-section.

Figure 4.8 shows the voltages and currents at different points of the conductor and screen, for a cross-bonded cable energised at peak voltage by means of synchronised switching.

---

[8] Check Eq. (3.116) in Sect. 3.4 and compare the two matrices.

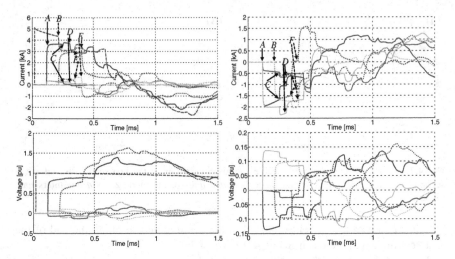

Fig. 4.8 Currents (*top*) and voltages (*bottom*) in a cross-bonded cable. *Left* conductor, *right* screen. *Dashed line* sending end, *solid line* end of the first minor-section, *dotted line* end of the second minor-section

The cable is energised and a wave propagates in cable, reaching the end of the first minor-section at Point *A* in Fig. 4.8. At this instant a current appears in all three screens, with a magnitude of ~ 2 kA for the screens of phases 1 and 3 and of ~ 400 A for the screen of phase 2.

The current in Phase 1 of the first minor-section generates a current in the screen with equal magnitude and opposite polarity, because of the coaxial mode as explained for the cable bonded in both-ends.

In the first crossing point, there is a transposition of the screens as shown in Fig. 4.9. The transposition of the screens results in a variation of the surge impedances, leading to reflections and refractions. If we consider the six modes to be fully decoupled, we can estimate the reflection and refraction of each mode. However, the modal impedance matrix depends on the frequency and an accurate estimation of the reflection/refraction coefficients requires knowing the frequency

Fig. 4.9 Currents in the conductors and screens for the first crossing point

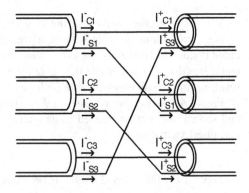

associated to the switching phenomena. Yet, the general tendency of the modes can still be attained.

The solving of the equations for different frequencies shows that the ground mode and the first coaxial mode are not affected by the cross of the screens and that the magnitudes of these two modes remain unchanged. The equations also show that the second and third coaxial modes are partly reflected with a negative polarity and refracted with a coefficient smaller than one, whereas the intersheath modes have a low reflection coefficient and a refraction coefficient slightly larger than 1.

At this point it is important to analyse the phase currents and see which information they can provide. Figure 4.9 shows the currents in the first cross-section. At this point, we know that prior to the cross there is only current in the conductor (I_{C1}^-) and screen (I_{S1}^-) of Phase 1, whereas the remaining currents are zero. We also know that the current of the first coaxial mode is not affected by the crossing of screens and that the current of the second and third coaxial modes decrease. Thus, we can write the equations for the first (4.1) and second (4.2) coaxial modes.

$$\frac{1}{\sqrt{3}}\left(I_{C1}^- + I_{C2}^- + I_{C3}^-\right) = \frac{1}{\sqrt{3}}\left(I_{C1}^+ + I_{C2}^+ + I_{C3}^+\right) \tag{4.1}$$

$$x\left(\frac{2}{\sqrt{6}}I_{C1}^- + \frac{1}{\sqrt{6}}\left(-I_{C2}^- - I_{C3}^-\right)\right) = \frac{2}{\sqrt{6}}I_{C1}^+ + \frac{1}{\sqrt{6}}\left(-I_{C2}^+ - I_{C3}^+\right), \text{ where } 0 < x < 1 \tag{4.2}$$

The magnitudes of the coaxial modes currents before the crossing are approximately 2.6, 3.5 kA and 0 A for, respectively, the first, second and third coaxial modes.[9] The refraction coefficient of the second coaxial mode[10] is approximately 0.7 and (4.1) and (4.2) can be manipulated into (4.3), resulting in a current I_{C1}^+ of approximately 3.5 kA.

$$\begin{cases} \frac{1}{\sqrt{3}}\left(I_{C1}^+ + I_{C2}^+ + I_{C3}^+\right) = 2.5 \times 10^3 \Leftrightarrow I_{C1}^+ + I_{C2}^+ + I_{C3}^+ = \sqrt{3} \cdot 2.5 \times 10^3 \\ \frac{2}{\sqrt{6}}I_{C1}^+ + \frac{1}{\sqrt{6}}\left(-I_{C2}^+ - I_{C3}^+\right) = 0.7 \cdot 3.5 \times 10^3 \end{cases} \tag{4.3}$$

The equations have also shown that after the crossing of the screens there will be a current in the conductor of Phases 2 and 3. Moreover, we know that these two currents are approximately equal, as there is neither transposition of the conductors nor third coaxial mode. Thus, I_{C2}^+ and I_{C3}^+ are approximately 400 A.

The summation of the magnitude of the three conductor currents after the crossing is equal to 4.3 kA, the same magnitude of the current before the crossing point.

---

[9] Equivalent to a current of approximately 4.3 kA in Phase A and 0 A in the other two phases.
[10] For a frequency of 1.5 kHz.

Having calculated the currents in the conductors it is time to calculate the currents in the screens. Using Fig. 4.9 as reference it is expected I_{S1}^+ and I_{S3}^+ to be equal in the crossing point. Consequently, the total screen current is composed of coaxial and intersheath modes currents, meaning that intersheath modes are generated at the crossing point.

Like with the conductors, the summation of the magnitude of all the three screen currents after the crossing point should be equal to the magnitude of all the screen currents before the crossing, in this case only I_{S1}^-. As a result, (4.4) can be written.

$$-4.3 \times 10^3 = I_{S1}^+ + I_{S2}^+ + I_{S3}^+ \Leftrightarrow -4.3 \times 10^3 = 2I_{S1}^+ + I_{S2}^+ \tag{4.4}$$

The current I_{S2}^+ can be estimated by knowing the current I_{C3}^+, as the intersheath mode currents should cancel for this phase and only the coaxial mode current remains. As a result, I_{S2}^+ is equal to -400 A, whereas I_{S1}^+ and I_{S3}^+ are equal to -1.9 kA.

The conductor and screen currents can be used to calculate the modal currents. Table 4.1 shows the currents in the conductor and screens for the different modes, results that are also shown in Fig. 4.10.[11]

The results are particularly interesting for the screens. As example, the total currents in screen 1 and screen 3 are the same at the crossing point, but the coaxial and intersheath mode currents are not. The coaxial mode currents propagate faster than the intersheath mode currents. Thus, the front wave of the currents in both screens is different, meaning that if we "measure" the current in both screens some meters ahead we see first the coaxial currents, $-3,500$ and -450 A for, respectively, screen 1 and screen 3 (in reality is now screen 2, but for simplicity we keep the nomenclature), and later the intersheath currents, $1,550$ and $-1,500$ A for, respectively, screen 1 and screen 3.

Table 4.1 Currents in the conductors and screens for the different modes

Mode	Total	Cond. 1	Cond. 2	Cond. 3	Screen 1	Screen 2	Screen 3
I_Coaxial_1	2,500	1,450	1,450	1,450	−1,450	−1,450	−1,450
I_Coaxial_2	2,500	2,050	−1,000	−1,000	−2,050	1,000	1,000
I_Coaxial_3	0	–	–	–	–	–	–
I_Intersheath_1	750	0	0	0	0	750	−750
I_Intersheath_2	1,400	0	0	0	1,550	−750	−750
I_Ground	0	–	–	–	–	–	–
Total	–	3,500	450	450	−1,950	−450	−1,950

The values shown in the table are not the magnitude of the modal currents, but how much of the current in each phase is generated by each mode

---

[11] The values are slightly different from the ones that would be obtained using the currents quoted previously. The correction is made to avoid confusion as there would be some small discrepancies between the theory and the values obtained, because of the approximations previously made in the calculation of some of the currents.

Fig. 4.10 Currents in the
conductors and screens
before and after the crossing
off the cables. In parenthesis,
the contribution for the phase
current of the coaxial and
intersheath modes,
respectively

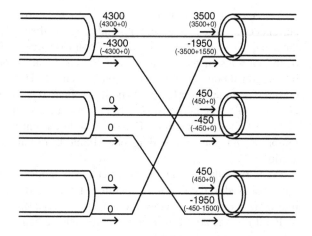

The current in the two screens have the same magnitude after the arrival of the intersheath mode currents, but between the arrival of the coaxial mode and intersheath mode currents, the currents in the two screens are different. We should keep this in mind for the analyses of the second cross-bonding point as the coaxial modes will arrived sooner than the intersheath modes.

After this analysis, we can observe again Fig. 4.8 and see if the theory fits in the simulation results.

Point A is the arrival of the coaxial currents to end of the first minor-section. Until this point everything is equal for the cross-bonded cable and the cable bonded in both-ends. At this instant, the current in screens of Phases 1 and 3 have a large magnitude (~ -2 kA), whereas the current in the screen of Phase 2 is ~ -400 A. All three current values are according to the theory,[12] the same happening for the currents in the screens whose amplitudes are 3.6 kA, 400 A and 400 A for, respectively, Phases 1, 2 and 3.

Point A shows also others differences between having a cross-boned cable or a cable bonded in both-ends, as there is a drop on the magnitude of the current in the conductor and screen of the cross-bonded cable when compared with the first minor-section. The drop is because of the reduction of the third coaxial mode that was previously explained. The same drop is seen when the coaxial currents reach the end of the second minor-section at Point B.

Point C is the arrival of the intersheath mode current generated at the energisation instant reaching the first minor-section. The current in the screen of Phase 1 receives a positive impulse, whereas the current in the other two screens receive an impulse with opposite polarity and half of the magnitude, similar to a cable bonded in both-ends.

---

[12] There are some small differences for Table 8, but it is normal when considering the several approximations that were made.

Yet, there is a difference in the conductor currents. Whereas the conductor currents of a cable bonded in both-ends are not affected by the intersheath current, the same is no longer valid for a cross-bonded cable, resulting in a small variation of the currents in the conductors of Phases 1 and 3, because of changes in the intersheath mode 2 and coaxial modes 2 and 3. This variation of the conductor currents is quite important as it typically increases the peak currents of a cross-bonded cable when compared with an equivalent cable bonded at both-ends.

Point D is the reflection of coaxial mode current reflected back in the second minor-section and the sending end (Point B instant) reaching the first minor-section.

Point E is the arrival of the intersheath modes generated when the coaxial modes arrived to the first minor-section (Point A) to the second minor-section and the sending end of the cable (the lengths are equal).

Point F is the arrival of coaxial mode reflected back in the receiving end of the cable to the second minor-section.

The remaining variations of the curves can be explained by similar process, but it becomes rather complicated because of the superimposition of so many waves.

Figure 4.11 shows the modal currents in first and second minor-sections, which confirm the behaviour of the waves that was previously exposed.

The voltage waveforms of Fig. 4.8 are also explained using the modal theory as done for the current waveforms. The voltage and current transformation matrices are different resulting in differences in the waveforms, but the variations are at precisely the same instants.

There are changes in the reflections are refractions coefficients: The refraction coefficient of the third coaxial mode increases and the refraction coefficients of two intersheath modes decrease; however, the first coaxial mode and the ground mode continue not being affected by the crossing of the screens. Like previously done, we can write the equations for the first (4.5) and second (4.6) coaxial modes for the point just before the crossing, knowing that there is only voltage at the conductor of phase 1 and the refraction coefficient for the second coaxial model. Ideally the magnitude of the voltage in phase 1 would be 1 pu just before the cross-point, however, we have seen in Figs. 4.2 and 4.4 that because of the non-linear elements

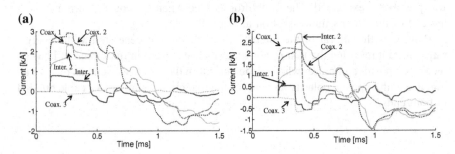

Fig. 4.11 Modal current during the energisation of the cable. **a** First minor-section. **b** second minor-section. *Solid lines* intersehat modes, *dashed lines* coaxial modes

of the cable it takes time for the voltage to reach that value, resulting in a lower voltage at the instant that the wave reaches the crossing point. In this specific case, the value is 0.85 pu.

$$\frac{1}{\sqrt{3}}\left(E_{C1}^- + E_{C2}^- + E_{C3}^- - \left(E_{S1}^- + E_{S2}^- + E_{S3}^-\right)\right) = \frac{1}{\sqrt{3}}\left(E_{C1}^+ + E_{C2}^+ + E_{C3}^+ - \left(E_{S1}^+ + E_{S2}^+ + E_{S3}^+\right)\right)$$

$$\frac{1}{\sqrt{3}}\left(E_{C1}^- + 0 + 0 - (0+0+0)\right) = \frac{1}{\sqrt{3}}(0.85) \simeq 0.5\,pu \tag{4.5}$$

$$x\frac{1}{\sqrt{6}}\left(2E_{C1}^- - E_{C2}^- - E_{C3}^- - 2E_{S1}^- + E_{S2}^- + E_{S3}^-\right) = \frac{1}{\sqrt{6}}\left(2E_{C1}^+ - E_{C2}^+ - E_{C3}^+ - 2E_{S1}^+ + E_{S2}^+ + E_{S3}^+\right)$$

$$1.25\frac{1}{\sqrt{6}}\left(2E_{C1}^- - 0 - 0 - 2\cdot 0 + 0 + 0\right) = 1.25\frac{1}{\sqrt{6}}(2\cdot 0.85 - 0 - 0 - 2\cdot 0 + 0 + 0) = 0.85\,pu \tag{4.6}$$

We know from the cable bonded in both-ends example that there is no voltage in the screens up to the end of the first minor-section.[13] Thus, the summation of all the screen voltage must continue to be zero after the crossing of the screens. At the same time, we know that $E_{S1}^+ = -E_{S3}^+$, as these are the two phases connected to the energised phase. Consequently, E_{S2}^+ is zero and the other two screen voltages have to be calculated.

We leave the screen in stand-by for a moment and we focus back on the conductor voltages. The equation of the first coaxial mode can be written as (4.7). There is no transposition of the conductors and thus, $E_{C2}^+ = -E_{C3}^+$.[14] As a result, we know that E_{C1}^+ is approximately 0.85 pu.

The voltage in the screen of Phase 1 can be calculated using the second coaxial mode as shown in (4.8). The final value of E_{S1}^+ being approximately -0.15 pu, whereas E_{S3}^+ is approximately 0.15 pu.

The magnitude of E_{C2}^+ and E_{C3}^+ is calculated using the third coaxial, which is zero, mode as shown in (4.9). The magnitudes are approximately -0.05 and 0.05 pu, respectively.

$$\left(E_{C1}^+ + E_{C2}^+ + E_{C3}^+\right) = 0.5\sqrt{3} \tag{4.7}$$

$$\left(2E_{C1}^+ - E_{C2}^+ - E_{C3}^+ - 2E_{S1}^+ - E_{S3}^+\right) = 0.85\sqrt{6} \Leftrightarrow 3\left(E_{C1}^+ - E_{S1}^+\right)$$
$$= 0.85\sqrt{6} + 0.5\sqrt{3} \tag{4.8}$$

$$E_{C2}^+ - E_{C3}^+ + E_{s3}^+ = 0 \tag{4.9}$$

---

[13] There is some voltage from the intersheath mode, but it has a lower velocity and it arrives later.

[14] The same conclusion can be achieved, by knowing that the third coaxial mode is 0.

Table 4.2 Voltages, in pu, in the conductors and screens for the different modes

Mode	Total	Cond. 1	Cond. 2	Cond. 3	Screen 1	Screen 2	Screen 3
E_Coaxial_1	0.5	0.3	0.3	0.3	0	0	0
E_Coaxial_2	0.85	0.7	−0.35	−0.35	0	0	0
E_Coaxial_3	0	–	–	–	–	–	–
E_Intersheath_1	−0.15	0	−0.08	0.08	0	−0.08	0.08
E_Intersheath_2	−0.25	−0.15	0.08	0.08	−0.15	0.08	0.08
E_Ground	0	–	–	–	–	–	–
Total	–	0.85	−0.05	0.05	−0.15	0	0.15

Table 4.2 shows the voltages, in pu, in the conductor and screens for the different modes. Like for the current, the values of the last row are slightly corrected to compensate for all the approximations that would result in some small contradictions with the theory.

We have seen that the coaxial modes generate two intersheath modes at the crossing points. We have previously seen for a cable bonded in both-ends that the intersheath mode affects the voltage in the core. As a result, a cross-bonded has typically a larger peak voltage than a cable bonded in both-ends. This is not a general rule, as the peak voltage depends on the length and number of minor-sections and in some exceptional cases it can even be lower.[15]

Table 4.3 shows the peak voltage values during the energisation for different bonding configurations. It is seen that the magnitude of the peak voltage is larger for a cross-bonded cable than for a cable bonded in both-ends. It may seem from this example that there is not a big difference between having one or more major-sections, as the variation of the peak value is relatively small. Yet, we will see later that in some situations the difference may be rather large.

The peak overvoltage instant can also be latter than for a cable bonded on both-ends. We have seen that the coaxial modes voltages are attenuated in the cross points. As a result, the voltage is less reduced when the coaxial wave reaches the receiving end by the second time and it is possible that the intersheath mode voltages arriving after this instant increase the voltage for values larger than those obtained before.

Table 4.3 Peak overvoltage for different bonding configurations

Both-ends	1.67
1 major-section	1.82
2 major-sections	1.80
3 major-sections	1.85
4 major-sections	1.86
5 major-sections	1.81

---

[15] An analysis similar to the one done before for the first crossing point would be very complicated because of the large number of superimposed waves. However, the readers can use the files available online to simulate this case if desired.

Summary

This section introduced to the importance of the bonding configuration and how it influences the waveforms. The analysis was made using the modal theory, which is necessary to explain the variations in the waveforms. The explanations given may seem quite confusing when read for the first time, but the readers can download the simulation files from the online page and use them to a step-by-step study.

An important result is the influence of the intersheath modes in the results. An intersheath mode voltage affects the voltage in the conductor, whereas an intersheath mode current does not. By other hand, a coaxial mode current influences the current in the screen, whereas the coaxial mode voltage does not.

The peak voltage is typically larger for a cross-bonded cable than for a cable bonded in both-ends and the precise magnitude depended on the length of the minor-sections and their number. For this particular example, there was not a big difference between having one or more major-sections, but we will see that for some situations the difference may be of several pu.

4.4 Energisation of Cables in Parallel

We have seen in Sect. 4.3 how it processes the energisation of an isolated cable. However, cables are normally installed in a network together with other cables and in some situation cables are energised in parallel, i.e., there are energised cable(s) connected to the sending end (busbar) of the cable being energised. In this situation, the energisation phenomenon becomes more complex as there is an interaction between the cables.

Electrically, cables are very similar to capacitors. As a result, the energisation of a cable, when connected to an already energised cable, can be seen as similar to the energisation of capacitor banks in parallel. The energisation of capacitor banks in parallel originates inrush currents whose amplitude and frequency depend on the voltage at the connection instant. More important, the frequency and magnitude of the inrush current associated to the energisation of the second capacitor bank are high when compared with the energisation of the first capacitor bank.

The cable capacitance is distributed uniformly along the cable. Thus, the energisation of cables in parallel does not result in inrush currents as large as those associated to the energisation of capacitor banks in parallel.

However, the phenomenon is still relevant and in some cases dangerous to the circuit breaker. A circuit breaker has a maximum tolerable amplitude and frequency for the inrush currents, defined by the IEC standard 62271-100. The surpass of this limit results in a wearing of the arcing contacts to a cone like shape and consequent degradation of the CB ability to interrupt short circuit currents and an increasing number of pre-strikes.

Figure 4.12 shows an example of the phenomenon, using an infinite series of RLC circuit for the cable modelling. When the CB is switched-on, the energy

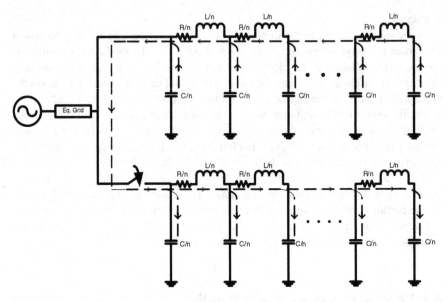

Fig. 4.12   Equivalent circuit for the energisation of cables in parallel

stored in the "capacitors" of the already energised cable is transferred to the "capacitors" of the cable being energised, as indicated by the arrows.

The phenomenon is also strongly dependent on the short-circuit power level at the sending end of the cable.

As an example, a very strong grid, i.e., infinite short-circuit power, is unaffected by switching operations and the phenomenon does not occur. In this situation, the voltage at the sending end is at all times the sinusoidal wave imposed by the system, which provides also all the energy to the cable being energised. In other words, if the cables in parallel are of the same type and have the same length, the energisation transients will be equal in both when connected to an ideal voltage source, as no energy is transferred between the cables.

By other hand, a very weak grid is strongly affected by any switching operations in the network. The feeding network of a weak grid has a slow response and the majority of the energy transferred to the cable being energised comes from the already energised cable(s) installed in parallel. In the limit case, almost all the inrush current energy is transferred between cables and there is both a high inrush current frequency and magnitude.

Another configuration that can lead to this phenomenon is to have both cables' sending ends connected to a busbar where the only other connection is a transformer. In this situation the transformer acts as large inductance, which for the effect delays the response from the feeding network leading to having the majority of the initial energy to be provided by the cable already energised.

Figure 4.13 shows an example of the energisation of a cable (*Cable B*) for peak voltage in parallel to an already energised cable (*Cable A*) of equal length in a very

Fig. 4.13 Current waves during the energisation of Cable B. *Solid line* current in cable A, *dashed line* current in cable B

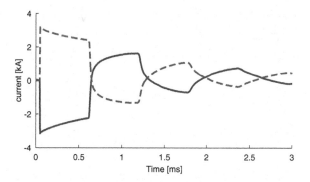

weak network. The currents in both cables are symmetrical during the first instants of the energisation, meaning that all the energy is being transferred from *Cable A* to *Cable B*.

4.4.1 Estimation Formulas

The estimation of the magnitude of frequency of the inrush current for the worst-case scenario can be made using direct formulas,[16] which are demonstrated in the next pages. This section has also the advantageous of allowing the readers to learn how to do more complex analytical analysis of cable systems without the aid of software packages.

The formulas have some simplifications, being the most notable to consider that no energy is provided by the source/feeding network and that all inrush current is transferred from *Cable A* to *Cable B* i.e., they are only valid for weak networks.

The current is calculated using (4.10), where: U_m is the crest of the applied voltage, U_t the voltage trapped in the cable,[17] Z_1 and Z_2 the cables surge impedances. However, the surge impedances need to be corrected for the frequency, as it is explained in the next section.

$$i_{1\,peak} = \frac{U_m - U_t}{Z_1 + Z_2} \qquad (4.10)$$

The calculation of the frequency of the inrush current is a more complicated process. For its calculation the cables are seen as two lossless pi-models in parallel (Fig. 4.14). This approach has the merit of allowing to analytical solve the system at the expense of a limited error in the final result, resultant of the simplifications

---

[16] This is an advantageous for many companies that do not have any advance simulation software, but still need to know if there is a risk for their system when energising cables in parallel.

[17] Typically the cable is discharged at the energisation instant and the value is 0 V.

Fig. 4.14 System used to
calculate the inrush current
frequency

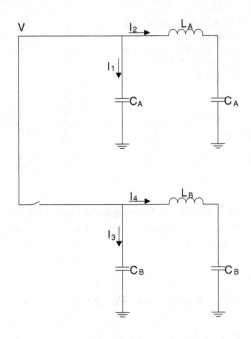

in the model: no resistance and a limited number of sections. However, the comparison of the formula with software simulations shows that the error is acceptable.

The currents are calculated as shown in (4.11). The second cable is considered as being unloaded; thus, all initial conditions in the cable are equal to zero.

$$\begin{cases} I_1 = sC_A V - C_A V(0) \\ I_2 = \frac{sC_A V - C_A V(0) + sL_A C_A I_2(0) - L_A C_A \dot{I}_2(0)}{s^2 L_A C_A + 1} \\ I_3 = sC_B V \\ I_4 = \frac{sC_B V}{s^2 L_B C_B + 1} \end{cases} \tag{4.11}$$

The value $V(0)$ is not precisely the same in I_1 and I_2, as the voltage does not have precisely the same magnitude along all the cable. However, in a well design system the difference should be minor and thus the values are considered to be equal.

$V(0)$ is considered as equal to the peak voltage and as the cable is mostly capacitive $I_2(0)$ is equal to zero and $\dot{I}_2(0)$ is approximately equal to the peak current.

Continuing to developing (4.11), (4.12) is obtained.

$$V\left(sC_A + \frac{sC_A}{s^2 L_A C_A + 1} + sC_B + \frac{sC_B}{s^2 L_B C_B + 1}\right) = C_A V(0) + \frac{L_A C_A \dot{I}_2(0)}{s^2 L_A C_A + 1} \tag{4.12}$$

Cables of equal type and length
If the cables are equal, L_A is equal to L_B and C_A to C_B, and (4.12) is written as (4.13), where $C = C_A = C_B$ and $L = L_A = L_B$. The current into the cable is then given by (4.14).

$$V = \frac{s^2 LC^2 V(0) + CV(0) + LC\dot{i}_2(0)}{s(s^2 2LC^2 + 4C)} \tag{4.13}$$

$$I = \frac{LC\dot{i}_2(0)}{s^2 2CL + 2} + \frac{CV(0)}{2} \tag{4.14}$$

Equations (4.15) and (4.16) are obtained when applying the inverse Laplace transform into (4.14) (where C is now the total cable capacitance).

$$I(t) = \frac{\sin\left(\sqrt{\frac{1}{LC}}t\right)\sqrt{LC}\dot{i}_2(0)}{2} + \frac{CV(0)\delta(t)}{2} \tag{4.15}$$

$$\omega_r = \sqrt{\frac{2}{LC}} \tag{4.16}$$

Cables of different length
To calculate the resonance frequency, the capacitance and inductance of the first cable to be energised are written in function of the capacitance and inductance of the second cable to be energised in (4.17) and (4.18). As example, if the Cable A capacitance and inductance are, respectively, 5 µF and 6.5 mH, whereas in *Cable B* are 3 µF and 5.5 mH, the value of x and y are, respectively, 0.6 and 0.846.

$$C_B = xC_A \wedge C = C_A \tag{4.17}$$

$$L_B = yL_A \wedge L = L_A \tag{4.18}$$

Equation (4.12) is written as (4.19), and the current is equal to (4.20).

$$V = \frac{s^4(yxL^2C^3V(0)) + s^2(yxLC^2V(0) + LC^2V(0) + yxL^2C^2\dot{i}_2(0)) + CV(0) + LC\dot{i}_2(0)}{s(s^4(yx^2L^2C^3 + yxL^2C^3) + s^2(yx^2LC^2 + 2xLC^2 + 2yxLC^2 + LC^2) + (2xC + 2C))} \tag{4.19}$$

$$I = \frac{C(s^4(yxL^2C^3V(0)) + s^2(yxLC^2V(0) + LC^2V(0) + yxL^2C^2\dot{i}_2(0)) + CV(0) + LC\dot{i}_2(0))}{s^4(yx^2L^2C^3 + yxL^2C^3) + s^2(yx^2LC^2 + 2xLC^2 + 2yxLC^2 + LC^2) + (2xC + 2C)}$$
$$+ \frac{C(s^4(yxL^2C^3V(0)) + s^2(yxLC^2V(0) + LC^2V(0) + yxL^2C^2\dot{i}_2(0)) + CV(0) + LC\dot{i}_2(0))}{(s^2LC + 1)(s^4(yx^2L^2C^3 + yxL^2C^3) + s^2(yx^2LC^2 + 2xLC^2 + 2yxLC^2 + LC^2) + (2xC + 2C))} \tag{4.20}$$

The partial fractions method cannot be applied to (4.20), as the numerator and denominator of the first term have the same order of the magnitude (s^4). Applying the propriety of Linearity, (4.20) is separated into two terms. This first term represents (4.21) in the frequency domain. Therefore, the frequency of I_3 is equal to the frequency of V, which can be calculated from (4.19).

$$I_3 = C_B \frac{dV}{dt} \tag{4.21}$$

The method of partial fractions is applied to the second part of the polynomial, which represents the current I_4, (4.22).

$$V = L_B \frac{dI_4}{dt} + \frac{1}{C_B} \int I_4 dt \tag{4.22}$$

The roots are conjugated complexes (4.23). Thus, the second part of (4.20) can be written as (4.24).

$$\begin{cases} \alpha_1 = \pm\sqrt{-\frac{1}{LC}} \\ \alpha_2 = \pm\sqrt{-\frac{yx^2+2yx+2x+1-\sqrt{y^2x^4+4y^2x^3-4yx^3+4y^2x^2-6yx^2+4x^2-4yx+4x+1}}{2CL_{(yx^2+yx)}}} \\ \alpha_3 = \pm\sqrt{-\frac{yx^2+2yx+2x+1+\sqrt{y^2x^4+4y^2x^3-4yx^3+4y^2x^2-6yx^2+4x^2-4yx+4x+1}}{2CL_{(yx^2+yx)}}} \end{cases} \tag{4.23}$$

$$I = \frac{A1s+A2}{(s-\alpha_1)(s-\alpha_1^*)} + \frac{A3s+A4}{(s-\alpha_2)(s-\alpha_2^*)} + \frac{A5s+A6}{(s-\alpha_3)(s-\alpha_3^*)} \tag{4.24}$$

Solving (4.24) it is obtained (4.25).

$$\begin{cases} A1 = 0 \\ A2 = yx^2L^3C^4\dot{i}_2(0) + yxL^3C^4\dot{i}_2(0) \\ A3 = 0 \\ A4 = yxL^2C^3\left[\frac{K_1\cdot(V(0)-L\dot{i}_2(0)-xL\dot{i}_2(0))+K_2}{K_3}\right] \\ A5 = 0 \\ A6 = -yxL^2C^3\left[\frac{K_1\cdot(-V(0)+L\dot{i}_2(0)+xL\dot{i}_2(0))+yx^3L\dot{i}_2(0)+K_2}{K_3}\right] \end{cases} \tag{4.25}$$

where K_1, K_2 and K_3 are given by (4.26).

$$\begin{cases} K_1 = \sqrt{y^2x^4 + 4y^2x^3 - 4yx^3 + 4y^2x^2 - 6yx^2 + 4x^2 - 4yx + 4x + 1} \\ K_2 = yx^3L\dot{i}_2(0) + yx^2(3L\dot{i}_2(0) - V(0)) + 2yx(L\dot{i}_2(0) - V(0)) \\ \qquad - 2x^2L\dot{i}_2(0) - 3xL\dot{i}_2(0) + V(0) - L\dot{i}_2(0) \\ K_3 = 2\sqrt{y^2x^4 + 4y^2x^3 - 4yx^3 + 4y^2x^2 - 6yx^2 + 4x^2 - 4yx + 4x + 1} \end{cases} \tag{4.26}$$

Thus, the resonance frequencies are calculated by (4.27) (where C is now the total capacitance of the cable). The frequencies of (4.20) first term are equal to (4.27).

$$\omega_{r1,2}^2 = \frac{(yx^2 + 2yx + 2x + 1) \pm \sqrt{y^2x^4 + 4y^2x^3 - 4yx^3 + 4y^2x^2 - 6yx^2 + 4x^2 - 4yx + 4x + 1}}{CL(yx^2 + yx)}$$

$$(4.27)$$

The resonance frequency correspondent to *A2* term is not considered because the term A2 is, for normal values of capacitance and inductance, several thousands of times smaller than A4 and A6.

If the two cables are of the same type, y is equal to x and (4.27) is simplified to (4.28) (where C is now the total capacitance of the cable).

$$\omega_{r1,2}^2 = \frac{(x^3 + 2x^2 + 2x + 1) \pm \sqrt{x^6 + 4x^5 - 6x^3 + 4x + 1}}{CL(x^3 + x^2)}$$

$$(4.28)$$

In summary, the frequency of the inrush current is calculated by:

- $\omega_r = \sqrt{\frac{2}{LC}}$, if the two cables have the same characteristics and lengths.
- $\omega_r^2 = \frac{(x^3+2x^2+2x+1) - \sqrt{x^6+4x^5-6x^3+4x+1}}{CL(x^3+x^2)}$, if the two cables have the same characteristics, but different lengths.
- $\omega_r^2 = \frac{(yx^2+2yx+2x+1) - \sqrt{y^2x^4+4y^2x^3-4yx^3+4y^2x^2-6yx^2+4x^2-4yx+4x+1}}{CL(yx^2+yx)}$, if the two cables have different characteristics.

4.4.2 Adjustment of the Inductance for High Frequencies

The frequencies associated to an inrush current are in order of hundreds of Hz or even kHz. At these high frequencies, the resistance and inductance of the cable are no longer the values given for 50/60 Hz given in datasheet.

We have seen in Sect. 3.3 how to calculate these parameters for a cable and we know that resistance increases with the frequency, whereas the inductance decreases. Thus, the inductance has to be corrected when calculating the frequency of the inrush current or else the calculated value is lower than the real one.

An exact calculation of the inductance value in function of the frequency requires the use of Bessel equations and the calculation of the *series impedance matrix*, as explained in Sect. 3.3.

The calculation of the inductance by this method is a rather complex process that many would prefer to avoid if possible, especially in the initial phase of a project. Table 4.4 presents some empirical formulas that can be used to estimate the inductance in function of the frequency for a single-core XLPE-type cable.[18]

---

[18] The authors want strongly to point out that these formulas are for a general cable, and will change depending on the cable characteristics and installation layout.

Table 4.4 Empirical formulas for the calculation of the inductance (H/km)

Cross-section (mm^2)	Copper—66 kV
800	$L(f) = 9.93 \times 10^{-4} f^{-0.6398} + 0.93 \times 10^{-4}$
1,200	$L(f) = 8.88 \times 10^{-4} f^{-0.6605} + 0.77 \times 10^{-4}$
2,000	$L(f) = 4.68 \times 10^{-4} f^{-0.5628} + 0.60 \times 10^{-4}$
Cross-section (mm^2)	Aluminium— 66 kV
800	$L(f) = 5.62 \times 10^{-4} f^{-0.4741} + 0.88 \times 10^{-4}$
1,200	$L(f) = 6.48 \times 10^{-4} f^{-0.5483} + 0.75 \times 10^{-4}$
2,000	$L(f) = 6.61 \times 10^{-4} f^{-0.6071} + 0.62 \times 10^{-4}$
Cross-section (mm^2)	Copper—150 kV
800	$L(f) = 2.83 \times 10^{-4} f^{-0.3358} + 1.31 \times 10^{-4}$
1,200	$L(f) = 2.63 \times 10^{-4} f^{-0.3633} + 1.12 \times 10^{-4}$
2,000	$L(f) = 2.08 \times 10^{-4} f^{-0.3601} + 0.93 \times 10^{-4}$
Cross-section (mm^2)	Aluminium—150 kV
800	$L(f) = 2.42 \times 10^{-4} f^{-0.2307} + 1.14 \times 10^{-4}$
1,200	$L(f) = 2.66 \times 10^{-4} f^{-0.3162} + 1.06 \times 10^{-4}$
2,000	$L(f) = 2.56 \times 10^{-4} f^{-0.3662} + 0.92 \times 10^{-4}$
Cross-section (mm^2)	Copper—400 kV
800	$L(f) = 2.70 \times 10^{-4} f^{-0.33} + 1.95 \times 10^{-4}$
1,200	$L(f) = 2.61 \times 10^{-4} f^{-0.3611} + 1.57 \times 10^{-4}$
2,000	$L(f) = 2.12 \times 10^{-4} f^{-0.3616} + 1.33 \times 10^{-4}$
Cross-section (mm^2)	Aluminium—400 kV
800	$L(f) = 2.34 \times 10^{-4} f^{-0.2268} + 1.78 \times 10^{-4}$
1,200	$L(f) = 2.65 \times 10^{-4} f^{-0.3159} + 1.52 \times 10^{-4}$
2,000	$L(f) = 2.59 \times 10^{-4} f^{-0.3663} + 1.32 \times 10^{-4}$

However, when calculating the frequency of the inrush current for the first time we have no idea of its value and we cannot therefore adjust the inductance. The solution is to use an iterative process where the frequency is first calculated for 50/60 Hz. After, the inductance is adjusted to the new value and the process continues until the variation stop.

To demonstrate the formula it is prepared a test-case consisting in two cables with different characteristics and lengths energised in a very weak network. The cable already energised (*Cable A*) is a 150 kV copper cable with a 800 mm^2 cross-section and 50 km length, whereas the cable being energised (*Cable B*) is a 150 kV copper cable with a 2,000 mm^2 cross-section whose length varies from 5 to 100 km in steps of 5 km.

Figure 4.15 shows the frequency of the inrush current in function of the length of *Cable B* for four different types of calculation methods. The first method is a simulation done in PSCAD that serves as reference. The second method is a calculation made using the proposed formula and with the correction of the inductance made using the equations of Sect. 3.3. The third method is analogous to the second, but the inductance is corrected using the empirical formulas of Table 4.4. The fourth and last method is similar to the previous two, but the

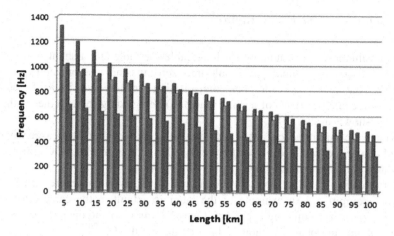

Fig. 4.15 Frequency of the inrush current for different cable lengths. *First column* PSCAD simulation, *second column* proposed formula with correction of the inductance, *third column* proposed formula with empirical correction of the inductance, *fourth column* proposed formula without correction of the inductance

inductance is not corrected for a high frequency and it is instead used as the datasheet value.

The comparison of the results shows that as long as the inductance is corrected for the frequency, the formula used for the estimation of the frequency is very accurate. The same does not happen if the inductance is left unchanged and in that case the error is in the order of hundreds of Hertz.[19]

There seems to be a larger error for the limit cases where the length of the *Cable B* is several times smaller than the length of *Cable A*. The difference is explained by the difficulties in obtaining a precise frequency for the PSCAD simulation of these cases, a consequence of the unusual waveforms present when a cable is several times longer than the other. Thus, it cannot be concluded that the error is in the formula, as the accuracy in the calculation of the frequency associated to the PSCAD simulation is lower.

The correction of the inductance must also be applied when estimating the magnitude of the inrush current (4.10) as the characteristic impedances depend on the frequency.

The peak inrush current for the simulation in PSCAD is 3,615 A. The estimation of the peak current according to (4.10) gives 2,370 A if the inductance is not corrected and 3,583–3,691 A if the inductance is corrected for the right frequency. Again, the correction of the inductance provides more accurate results.

---

[19] This also serves to re-demonstrate on how important it is to correct the electrical parameters for the right frequency. Many times there are used the 50 Hz values as they are the only ones available in the datasheet. However, for more than one phenomenon this simplification will lead to inaccurate results.

4.5 Zero-Missing Phenomenon

In some situations, most notably for long cables, a cable and a shunt reactor are energised together, which may result into zero-missing phenomenon. A phenomenon that may only occur when compensating for more than 50 % of reactive power of a cable by means of a shunt reactor(s) being energised together with the cable as shown in Fig. 4.16.

When zero-missing occurs, it is not possible to open the circuit breaker without risk of damage, except if the circuit breaker is designed to interrupt DC currents or open at a non-zero current value.

Zero-missing can last several seconds and it represents a severe risk to the network equipment. An example of the latter is a fault in the cable during its energisation, if zero-missing is present it may be impossible to open the poles of a CB in the healthy phases without risking damaging the CB.[20]

An easy way of understanding the zero-missing phenomenon is by analysing an inductor (equivalent to a shunt reactor) in parallel with a capacitor (equivalent to a cable) of equal impedance. In this situation, the currents in the capacitor and inductor have equal amplitude and are in phase opposition. The current in the inductor can also have a DC component, whose value depends on the voltage at the moment of connection.

In an inductor there is a 90° phase difference between the current and the voltage at its terminals. Thus, if the voltage is zero, the current should be maximum and vice versa. Due to energy conservation, the current in an inductor is continuous with the value zero prior to energisation, and so it must also be zero after the connection regardless of the voltage at the moment of connection. Therefore, if the inductor is connected for zero voltage in order to maintain its continuity, the current will have a DC component with an amplitude equal to the amplitude of the AC component. If the inductor is connected for a peak voltage, no DC component is present.[21]

Thus, if the LC circuit is energised when the voltage is crossing zero the DC current component in the inductor is maximum, the inductive and capacitive AC current components cancel each outer out and the current in the CB contains only the decaying DC component. In this situation it is impossible to open the CB as the current does not cross zero.

If the CB closes when the voltage is a peak value there is no DC current component and consequently no zero-missing phenomenon and the current in the CB is 0 V. For the cases between 0 V and peak voltage, the existence of zero-missing phenomenon depends on the switching angle.

---

[20] This may not seem an issue in some situation as the faulty phase would still be open when using a CB with synchronise switching. Yet, to have one phase open and the other two close may lead to ferroresonance and a long temporary overvoltage that is even more undesired.

[21] See Chap. 2 for more information on the energisation of inductive loads.

Fig. 4.16 Single-line diagram of a system with cable, shunt reactor and circuit breaker

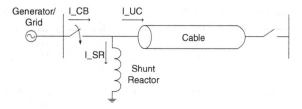

Substituting the inductor by a shunt reactor and the capacitor by a cable and it is obtained a similar behaviour, but also with damping because of the system resistance. In this situation, besides the switching angle the compensation level also influences the existence of the phenomenon as it determines the magnitude of the resultant AC current in the CB.

Figure 4.17 shows an example of zero-missing phenomenon for a 150 kV, 50 km cable, with 70 % of the reactive power being compensated by a shunt reactor installed at the cable sending end. The system is energised when the voltage is zero, and therefore the value of the DC component is at maximum, whereas the magnitude of the AC current in the CB is equal to 43 % of the magnitude of the current in the shunt reactor. In this situation, the DC current is larger than the resultant AC current and there is zero-missing phenomenon.

Figure 4.17a shows the currents in the CB (solid line), in the shunt reactor (dashed line) and in the cable (dotted line). It should be noted that the current in the CB does not cross zero. Figure 4.17b shows the current in the CB for a period of 5 s, where it can be seen that it takes several seconds before the current crosses zero for the first time.

In a real system with the shunt reactor installed at the sending end of the cable the decay of the DC current component may take several seconds to disappear depending on the cable and shunt reactor resistances, and its initial value depends both on the voltage value in the shunt reactor's terminals and the reactive power compensation level, by other words the shunt reactor inductance.

In a system with an ideal voltage source and the shunt reactor installed at the cable sending end, the decay time constant depends mostly on the X/R ratio of the

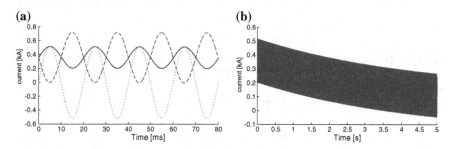

Fig. 4.17 Currents during the energisation of cable + shunt reactor system at zero voltage and 70 % reactive power compensation. *Solid line* current in the CB, *dashed line* current in the shunt reactor, *dotted line* Current in the cable

shunt reactor, which may vary quite a great deal from shunt reactor to shunt reactor. In the previous example, the decay time constant was 4.3 s. The time needed for the current in the CB to cross zero is estimated by (4.29), being approximately 3.6 s, value confirmed by the simulation.

$$t = -\frac{L_S}{R_S} \ln\left(\left|\frac{x-1}{x\cos\theta}\right|\right) \tag{4.29}$$

where: L_S is the shunt reactor inductance, R_S is the shunt reactor resistance, x is the reactive power compensation ratio and cos is the switching angle for a voltage defined by a cosines function, i.e., 0° equal to peak voltage with positive polarity, 90° equal to zero voltage and 180° equal to peak voltage with negative polarity.

The location of the shunt reactor is another factor influencing the duration of zero-missing phenomenon. The installation of the shunt reactor at the receiving end of the cable can substantially reduce the duration of the phenomenon. In this particular example, it would reduce the duration of the phenomenon from 3.6 to 0.63 s.

The reduction is explained by the changes in the current closing loop. If the shunt reactor is installed at the receiving end of the cable, the current closing loop has to pass by the cable, whose X/R ratio is hundred times smaller than the one of the shunt reactor. Consequently, there is an increase of the resistance of the loop and a faster damping of the transient DC current component.

Figure 4.18 shows the DC current loop for a shunt reactor installed at the receiving end of cable energised by an ideal voltage source. The capacitance of the cable is like an open circuit for the DC current and the only available current loop consists in the current closing through the ideal voltage source and the cable series impedance.

In this situation it is necessary to change the formula (4.29) to include the cable's resistance and inductance and also the equivalent network impedance. Thus, the worst case scenario is when the shunt reactor is installed at the sending end of the cable.

It seems at first that when compensating for more than 50 % of the cable reactive power, the network operator has to decide between having switching overvoltages or zero-missing phenomenon (assuming that the CB has synchronise

Fig. 4.18 DC current loop for a shunt reactor installed at the receiving end of the cable

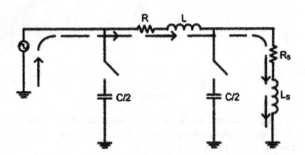

switching and the operator controls the switching instant, otherwise the switching is random and it is not possible to "choose").

In this situation it will normally be chosen to have zero-missing phenomenon as it is less hazardous to the network. Thus, countermeasures are necessary.

4.5.1 Countermeasures

The simplest countermeasures used to avoid zero-missing phenomenon are immediately deduced. If the shunt reactor is connected to the busbar or the cable through a CB, it is possible to control its closing time and energise the shunt reactor for peak voltages, avoiding DC currents. The drawback of this counter-measure is that it is necessary to buy an extra CB, which in order to be 100 % efficient should have synchronised switching capability, thus increasing the financial costs. Thus, it is beneficial to analyse other countermeasures.

Different reactive power compensation levels and locations
To have zero-missing phenomenon it is necessary that the shunt reactor compensates for more than 50 % of the reactive power generated by the cable and is energised together with the cable.

The reactive power generated by a long cable is so high that very often the compensation can be made by more than one shunt reactor. When this happens, the shunt reactors can and should be installed at different locations, one possibility being to connect part of the reactive power compensation, less than 50 %, directly to the cable and the remaining to the busbar, which can be also used to control the network voltage.

This method is a possibility for meshed grids, where several cables are connected to the same busbar and it is necessary to use the shunt reactors to control the system voltage.

Use of a pre-insertion resistor
A pre-insertion resistor consists of resistor blocks connected in parallel with the circuit breaker's breaking chamber that typically closes the circuit 8–12 ms before the arcing contacts (in this analysis it is considered 10 ms).

The value of the pre-insertion resistor should be such that the DC component becomes very small (ideally null) after 10 ms, in other words, the DC component is damped in the first 10 ms.

To eliminate the DC component in the first 10 ms, the pre-insertion resistor value should be precise. If the pre-insertion resistor is too small, it will not be able to damp the entire DC component in just 10 ms.

If the pre-insertion resistor is too large, it will be equivalent to an open circuit. Thus, when the pre-insertion resistor is bypassed the transient is similar to one obtained if no pre-insertion resistor was used. As the pre-insertion resistor is connected during 10 ms, when it gets disconnected the voltage value of the generator is symmetric to the value at the time of connection ($V_1(10$ ms$) = -V_1(0$ ms$)$), and so the DC component will persist, but with opposite polarity.

Calculation of the pre-insertion resistor value

To calculate the pre-insertion resistor value a pi-model representation of the cable, as shown in Fig. 4.19, will be used. Where V is the voltage source, R_S the shunt reactor resistance; L_S the shunt reactor inductance, R the cable's series resistance, L the cable's series inductance, C half of the cable's shunt capacitance and R_P the pre-insertion resistor.

The system equations are shown in (4.30). As the system cannot be analytically solved, it is necessary to use of numerical methods.

$$\begin{cases} V_2 = L_s \frac{dI_s}{dt} + R_s I_s \\ V_2 = \frac{1}{C} \int I_c dt \\ V_2 = R \cdot I_3 + L \frac{dI_3}{dt} + \frac{1}{C} \int I_3 dt \\ I_1 = I_2 + I_3 + I_c \\ V_2 = V_1 \cos(\omega t) - R_p \cdot I_1 \end{cases} \tag{4.30}$$

However, it is possible to introduce some simplifications in order to obtain a first approximation of the pre-insertion resistor value. This is done by calculating the energy that the pre-insertion resistor should dissipate (4.31).

$$W = \frac{1}{2} L_s \left(I_s^{DC} \right)^2 \tag{4.31}$$

The energy dissipated in the pre-insertion resistor is calculated by the integral in (4.32), whose limits are the time during which the pre-insertion resistor is connected.

$$W = \int P dt \Leftrightarrow W = \int_0^{0.01} R_p I_1^2 dt \tag{4.32}$$

The objective is to calculate R_p, where both I_1 and I_s^{DC} depend on the connection instant and are unknown. For 100 % reactive power compensation, the AC components of the shunt reactor and cable's currents cancel each other out, and at the moment of connection the current I_1 is equal to I_s^{DC}, whereas both should ideally be zero after 10 ms.

Fig. 4.19 Equivalent scheme of the shunt reactor and the cable when using a pre-insertion resistor

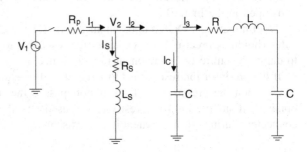

Considering that the current I_1 decreases linearly (this is an approximation, but as R_p is large the error is small), and neglecting R_S (which is much smaller than R_p), (4.32) can be simplified to (4.33), and the value of R_p is calculated by (4.35).

$$W = 0.01 R_p \left(\frac{I_1(0)}{2} \right)^2 \tag{4.33}$$

$$0.01 R_p \left(\frac{I_1(0)}{2} \right)^2 = \frac{1}{2} L_s (I_s^{DC})^2 \Leftrightarrow 0.01 R_p \left(\frac{1}{2} \right)^2 = \frac{1}{2} L_s \tag{4.34}$$

$$R_p = \frac{2 L_s}{0.01} \tag{4.35}$$

Because of the simplifications this method is not always accurate. If the DC component is maximum, the error can be disregarded, but if the DC component is smaller, i.e., the switching angle is not zero, the error increases.

The use of differential equations allows a more accurate calculation of the pre-insertion resistor value, but an iterative process is required to calculate the value of R_p. The program increases R_p, until it reaches a value at which the DC component is damped in 10 ms.

To verify that the DC component is damped, the peak value of I_S is calculated 10 ms after connection. For that value to be equal to the amplitude of the AC component, the DC component must be equal to zero. So when the calculated value is equal to (4.36) plus a small tolerance the iterative process stops. The value of V_2 depends on the value of the variable R_P and it is calculated in (4.37). The equation is solved using the partial fraction method and the inverse Laplace transform is applied as final step.

$$I_s^{peak} = \frac{V_2}{\sqrt{R_s^2 + (\omega L_s)^2}} \tag{4.36}$$

$$V_1(s) = V_2 \left(\frac{A}{N} + R_P \frac{B(1) + B(2)}{N} \right) \tag{4.37}$$

where:

$$A = s^5 LL_s C\omega^2 + s^4 \left(L_s RC\omega^2 + LCR_s\omega^2 \right) + s^3 \left(L_s\omega^2 + RCR_s\omega^2 + LL_s C\omega^4 \right)$$
$$+ s^2 \left(L_s RC\omega^4 + LCR_s\omega^4 + R_s\omega^2 \right) + s \left(L_s\omega^4 + RCR_s\omega^4 \right) + R_s\omega^4$$
$$N = s^3 LL_s C\omega^3 + s^2 \left(L_s RC\omega^3 + LCR_s\omega^3 \right) + s \left(L_s\omega^3 + RCR_s\omega^3 \right) + R_s\omega^3$$
$$B(1) = s^4 LC\omega^2 + s^3 RC\omega^2 + s^2 \left(\omega^2 + LC\omega^4 \right) + sRC\omega^4 + \omega^4$$
$$B(2) = s^6 LC^2 L_s\omega^2 + s^5 \left(LC^2 R_s\omega^2 + RL_s C^2\omega^2 \right) + s^4 \left(RR_s C^2\omega^2 + 2CL_s\omega^2 + LL_s C^2\omega^4 \right)$$
$$+ s^3 \left(2CR_s\omega^2 + LR_s C^2\omega^4 + RL_s C^2\omega^4 \right) + s^2 \left(RR_s C^2\omega^4 + 2CL_s\omega^4 \right) + s2R_s C\omega^4$$

Fig. 4.20 Current in the
circuit breaker and the shunt
reactor during the
energisation of the
cable + shunt reactor system
at 0 V. *Solid lines* with pre-
insertion resistor, *dashed
lines* without pre-insertion
resistor

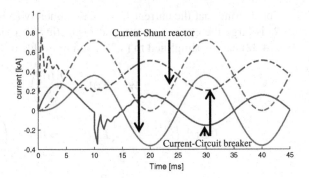

The value of the pre-insertion resistor depends on the initial value of the DC component. As this value depends on the connection moment, the equations are normally solved for the worst-case scenario, maximum DC current. For that case, the calculated value of R_p is ideal, whereas for the other cases the error is small. If desired the equations can be solved for other initial DC current values.

To check the efficiency of the pre-insertion resistor, we energise the 50 km long cable compensated at 70 % whose original energisation is shown in Fig. 4.17.

According to this method, the pre-insertion resistor should be 264 Ω. Figure 4.20 compares the energisation of the system with and without pre-insertion resistor, showing how the pre-insertion resistor completely damps the DC current in 10 ms, whereas without pre-insertion resistor it would take 3.6 s. The bypassing of the pre-insertion resistor is seen at 10 ms where it is observed a transient in the circuit breaker's current.

Even if the value of the pre-insertion resistor is not ideal, the DC component of the shunt reactor current is always reduced. Figure 4.21 shows the value of the DC

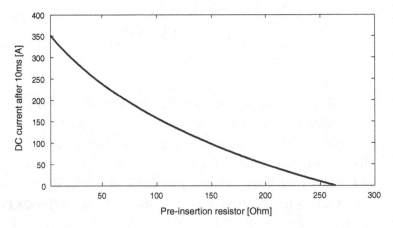

Fig. 4.21 Value of the DC component after bypassing the pre-insertion resistor, for phase closing when the voltage is zero and 100 % of reactive power compensation

component after 10 ms for different resistance values when the cable and the shunt reactor are connected for zero voltage.

The curve in Fig. 4.21 is non-linear, and the DC component is very small for pre-insertion resistor values close to the ideal. However, for larger differences, zero-missing phenomenon may still persist during long periods of time.

If, for instance the value of the pre-insertion resistor value was calculated using the energy Eq. (4.35) instead of the differential equations, the initial DC current after 10 ms would be about 19.5 A, smaller than the AC current in the CB and there would not be zero-missing phenomenon.

As a matter of fact, for this precise example there is not zero-missing phenomenon for pre-insertion resistances larger than 103 Ω (assuming resistance with the same order of magnitude), as the DC current after 10 ms is equal to the AC peak value and the current crosses zero. Yet, it will still take several cycles to fully damp the DC current.

However, as the reactive power compensation increases and the AC current decreases the more accuracy is necessary in the calculation of the pre-insertion resistance value. In the limit case of 100 % reactive power compensation, the value should be exact.

4.6 De-Energisation of a Cable

When a cable is disconnected the energy stored in the cable has to be dissipated which may take several seconds, because of the low cable resistance and high capacitance.

The disconnection of a cable is similar in many ways to the disconnection of a capacitor, and it is easier to understand if the latter is explained first. The current of an ideal capacitor leads the voltage in 90°. Thus, when a capacitor is disconnected, it is fully charged and has a voltage at its terminals of ±1 pu.[22] The capacitor's energy is then usually discharge through a resistance, but as the resistance value is usually low when compared with the capacitive reactance, a complete de-energisation of the capacitor can take a long time.

The de-energisation of a cable alone is alike the de-energisation of a capacitor, with the difference that the voltage after the disconnection is slightly smaller than 1 pu because of the resistance of the cable, and the damping is faster because of the higher resistance of a cable when compared with a capacitor bank. Thus, a typical de-energisation of a cable does not raise any possible problems to the network and it is only necessary to be careful in not to re-energise the cable immediately after the disconnection because of the stored energy.

In the previous section, we discussed the energisation of a cable and shunt reactor together. If they are energised together they are also de-energised together.

---

[22] Remember that a CB typically switch-off for 0 A if operating properly.

Due to this, the voltage is no longer a decaying DC component, but a decaying AC component oscillating at resonance frequency, whose approximate magnitude is given by (4.38). Where, L_S is the shunt reactor inductance, C the cable capacitance, R the cable resistance plus the shunt reactor resistance, $V_1(0)$ the voltage at the disconnection moment in the end without shunt reactor and $V_2(0)$ the voltage at the disconnection moment in the end connected to a shunt reactor.[23]

$$V = \frac{V_1(0) + V_2(0)}{2} \cdot \frac{\cos\left(\frac{1}{\sqrt{L_S C}}t\right)}{\exp\left(\frac{R}{8L_S}t\right)} \tag{4.38}$$

Equation (4.38) demonstrates that the resonance frequency is a function of the cable capacitance and the shunt reactor inductance, equalling 50 Hz if the shunt reactor compensates for all the reactive power generated by the cable. Usually the shunt reactor will not compensate for such a large amount of reactive power, and the resonance frequency is lower than 50 Hz, typical between 30 and 45 Hz.[24]

The voltage in (4.38) is a decaying sinusoidal, whose decaying ratio is inversely proportional to the reactive power provided to the shunt reactor and there is not an overvoltage associated to it.

If the power networks were single-phase the analysis would stop now, but as they are three-phase it is necessary to study the phenomenon a little more. When the cable is disconnected the three phases are disconnected in different moments, and there is normally a time difference of approximately 3.333 ms between the disconnections of each phase.[25] After the first phase is disconnected, the voltage and current in that phase start to oscillate at resonance frequency while the voltage and current in the other two phases continue to oscillate at system frequency. Therefore, the system is no longer balanced after the disconnection of all the three phases and the phase difference between phases is no longer 120°.

The analysis made until now had not yet taken the mutual coupling between phases into consideration. The mutual coupling between the phases of the cable will barely change the waveforms, but the same may not be true when considering the mutual coupling between the phases of the shunt reactor.[26]

We start by assuming that the mutual coupling is the same between all the phases, the shunt reactor compensates 70 % of the reactive power generated by the cable and the phase disconnection sequence is A–C–B.

Prior to the disconnection there is a 120° phase difference between the phases. Afterwards the disconnection of the first phase, Phase A starts to oscillate at 42 Hz while the other two phases continue to oscillate at 50 Hz. After 3.33 ms Phase C is

---

[23] $V1(0)$ and $V2(0)$ should be similar, the existence differences would be a result of Ferranti Effect.

[24] The resonance frequency can be also calculated by $f_r = f_N \cdot x$, where f_N is the system frequency and the x the reactive power compensation ratio.

[25] For a 50 Hz system, 2.778 ms for a 60 Hz system.

[26] For more information on shunt reactors including mutual coupling see Sect. 1.4.

disconnected. During these 3.33 ms Phases B and C rotate 60°, while Phase A rotates 50.4°. Thus, the difference between the phases 3.33 ms after the disconnection of phase A is: AB–110.4°; AC–129.6°; BC–120°.

After the disconnection of Phase C, Phases A and C oscillate at 42 Hz, while Phase B continues to oscillate at 50 Hz. After another 3.33 ms Phase B is disconnected. During these 3.33 ms Phase B rotates 60°, while Phases A and C rotate 50.4°. Thus, the angle between phases after the disconnection of all three phases is: AB–100.8°; AC–129.6°; BC–129.6°.

Table 4.5 shows the angle of each phase during the disconnection and Fig. 4.22 shows the phasors of all three phases and the voltage induced by mutual coupling after the disconnection of the three phases.

The phasor representation shows that for this specific case it is expected a larger voltage in Phase C, the second phase to be disconnected, than in the other two phases. Yet, this does mean that there is an overvoltage in Phase C as that depends on the value of the inductive coupling between the phases. We should keep in mind that the voltage starts to be damped at the disconnection instant and that the increase on the voltage because of mutual coupling may not be high enough to result in an overvoltage. Thus, if the coupling is small the voltage in Phase C is still larger than in the other two phases, but smaller than 1 pu.

Table 4.5 Position of the current vectors during the disconnection

	IA	IB	IC
Disconnection of Phase A	0°	−120°	120°
Disconnection of phase C (3.33 ms after the disconnection of Phase A)	50.4°	−60°	180°
Disconnection of phase B (3.33 ms after the disconnection of Phase C and 6.66 ms after the disconnection of Phase A)	100.8°	0°	230.4°

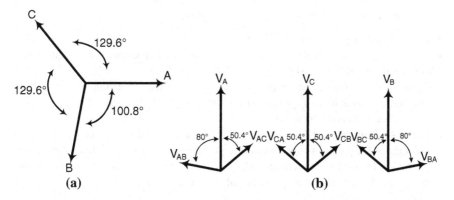

Fig. 4.22 a Current phasors after the disconnection of the three phases. **b** Self-voltage and voltage due to mutual coupling (not at scale) in all phases after the disconnection of the three phases

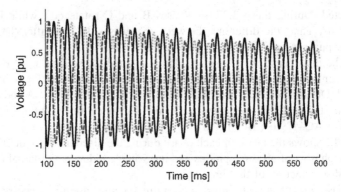

Fig. 4.23 Voltage at the end of the cable during the de-energisation of a cable + shunt reactor system. *Dotted line* phase A, *dashed line* Phase B, *solid line* phase C

Figure 4.23 shows the voltage at the receiving end of the cable during the de-energisation of the system previously described. To better observe the phenomenon the mutual coupling between phases was exaggerated and considered to be − 0.1 H between all phases.

The simulation shows a small overvoltage in Phase C, with a peak value of 1.09 pu, and what seems to be a low frequency component in all three phases. So, the next question is to understand from where the low frequency comes.

The mutual coupling between phases increases the complexity of the phenomenon's mathematical analysis and (4.38) is no longer valid. The full mathematical development of the equations is long and laborious, and it is not presented in this book.[27] Instead, there are presented the final results and a theoretical description of them.

As previously explained, the resonance frequency during de-energisation depends on the cable capacitance and shunt reactor inductance. The existence of mutual coupling between the phases adds two/three more frequencies to the wave, besides the resonance frequency, as given below:

- Cable de-energisation: decaying DC;
- Cable + shunt reactor de-energisation (no mutual inductance): Resonance frequency (typically < 50 Hz);
- Cable + shunt reactor de-energisation (equal mutual inductances): 2 mutual inductance frequencies (approximately given by (4.39), where M is the mutual inductance between phases);
- Cable + shunt reactor de-energisation (different mutual inductances): 3 mutual inductance frequencies (approximately given by (4.40), where M_{min} is the value of the lower mutual inductance, M_{max} is the value of the larger mutual inductance and M_{avg} is the average value of all mutual inductances).

---

[27] The full mathematical development is available at [3].

$$f_1 \approx \frac{1}{2\pi} \sqrt{\frac{1}{C(L_S - M)}}$$

$$f_2 \approx \frac{1}{2\pi} \sqrt{\frac{1}{C(L_S + 2M)}}$$
$$(4.39)$$

$$f_1 \approx \frac{1}{2\pi} \sqrt{\frac{1}{C(L_S - M_{\min})}}$$

$$f_2 \approx \frac{1}{2\pi} \sqrt{\frac{1}{C(L_S - M_{\max})}}$$
$$(4.40)$$

$$f_3 \approx \frac{1}{2\pi} \sqrt{\frac{1}{C(L_S + 2M_{\text{avg}})}}$$

That analysis of (4.39) shows that because of mutual coupling the resonance frequency is changed (f_1) and a second frequency is imposed (f_2). The low frequency component that seems to exist in Fig. 4.23 is in reality the difference between these two components.

The difference between f_1 and f_2 is only some Hertz. Thus, when both signals are superimposed on each other it gives the impression that there is a low frequency component. From mathematical point of view, it consists only in the addition of two sinusoidal waves with different frequencies.

If the waveform is decomposed in the frequency domain it is observed that the magnitude of the voltage for the frequency f_1 is several times larger than the magnitude of the voltage for the frequency f_2, explaining the low magnitude of what seems to be a low frequency component.

Another way of observing this phenomenon is to think in the disconnection of long OHL, where a similar phenomenon is observed because of the capacitive coupling between the phases.

Equation (4.40) shows that there are three different frequencies if the mutual coupling values are different. To be precise it is important to refer that mathematically there are also three frequencies when the mutual coupling is the same for all phases, but two of them are equal; it is easy to see in (4.40) that f_1 and f_2 are the same in a system where all mutual coupling values are equal.

It is difficult to point out the dominant frequency in this situation, as the difference between the magnitude of the voltage waves associated to f_1 and f_2 is not as larger as before. However, in a real system these two frequencies should be very close eliminating the problem.

The vectorial representation of the voltage induced by mutual coupling shown in Fig. 4.22 also changes if the mutual couplings between phases are different. The angles remain the same, but the magnitude of the induced voltages is not the same for all six. In this situation, we no longer can say that voltage in Phase C is larger than in the other two phases.

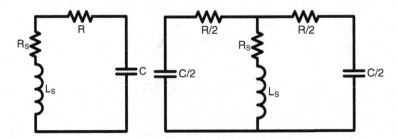

Fig. 4.24 Equivalent circuit for the cable and shunt reactor. **a** Shunt reactor installed at the sending end. **b** Shunt reactor installed in the middle of the cable

The location of the shunt reactor has also influence in the de-energisation waveform. If the shunt reactor is installed at the middle of the cable, the reactive power compensation divides itself into two equal parts as explained in Sect. 1.4. On other words, the magnitude of the inductive current in the cable is half of what would be if the shunt reactor was installed at one end of the cable.

If we think on the cable as being a resistor in series with a capacitor (the inductance of the cable is small when compare with the shunt reactor inductance), we can see that if the shunt reactor is installed in the middle of the cable there is a parallel between the RC load of each side (Fig. 4.24).

As a result, the resistance of the time constant of the de-energisation is lower and thus also the time constant, leading to a longer de-energisation when the shunt reactor is installed in the middle of the cable.

We have seen that an overvoltage may be created after the disconnection of a cable + shunt reactor system because of the mutual coupling between the phases of the shunt reactor. However, the example presented in this section shows that the overvoltage is small. Moreover, the example was for an inductive coupling larger than normal and the overvoltage in a real system is normally even smaller.

Thus, it is raised the question on why is this relevant from a practical point of view. The answer is: because it may influence the overvoltage associated to a restrike. But before going there, we need to understand what a transient recovery voltage (TRV) is.

4.7 Transient Recovery Voltage and Restrikes

We have seen that after the disconnection of a cable, the voltage in the cable is a decaying DC wave or a decaying AC wave depending if a cable is disconnected with or without a shunt reactor.

For simplicity, we start by explaining transient recovery voltages for a decaying DC voltage. In this case, the voltage at the cable sending end is ± 1 pu after the switch-off of the CB. The voltage in cable takes several seconds to be damped and it can be considered as constant and equal to ± 1 pu during the first instants of the de-energisation.

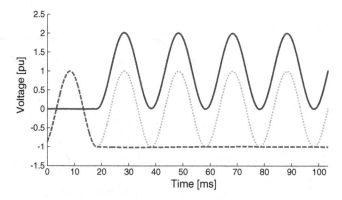

Fig. 4.25   Voltage immediately after the switch-off of the CB. *Solid line* CB TRV, *dashed line* voltage at the sending end of the cable, *dotted line* voltage at the source

In an ideal and instantaneous disconnection, the voltage at the source/network side after the switch-off of the CB continues to oscillate at power frequency with the same magnitude that it had before the opening of the CB.[28] Thus, the voltage at the CB terminals is a sinusoidal with a magnitude of 1 pu and an offset of ±1 pu, depending on the polarity of the voltage in the cable. This voltage at the CB terminals is called transient recovery voltage (TRV).

Figure 4.25 shows for one phase the voltage at the CB terminals, at the cable sending end and at the source side for a cable connected to an ideal voltage source. The figure shows that the voltage at the CB terminals increases after the switch-off reaching a peak value of 2 pu half-cycle (10 ms) later.

Until now we assumed an ideal and instantaneous disconnection of the CB, but the contacts of a CB do not separate immediately. During the time it takes for the contacts to separate, a restrike/reignition may occur.

The occurrence of a restrike/reignition depends on the CB dielectric strength. Figure 4.26 shows the withstand voltage of a general CB in function of time (it should be noted that this curve is linearised and it is an approximation to a real curve). As expected, the withstand voltage increases with the time as the CB contacts get further and further apart.

If the TRV exceeds the withstand voltage, a restrike/reignition is likely to occur. In the example shown in Fig. 4.26, the TRV is at one point higher than the withstand voltage and the system is re-energised at that instant. As the CB re-closing occurs in the first quarter of the cycle, the phenomenon is called reignition instead of restrike.

When a restrike/reignition occurs there is an arc at the CB contacts, resulting is a current flowing between the contacts, which lead to the equalisation of the

---

[28] Assuming a strong network. In a weak network the voltage would change when the cable is disconnected.

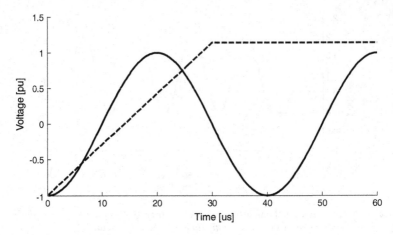

Fig. 4.26 Withstand voltage curve (*dashed line*) and system TRV (*solid line*)

voltage in both contacts. Consequently, it is initialised a transient alike to a switching transient.

Thus, the worst-case scenario is to have a voltage difference of 2 pu between the source's voltage and the cable's voltage, i.e., a restrike half a cycle after the CB opening. We saw in Sect. 4.3 that the peak magnitude of the switching overvoltage depends on the voltage difference at the CB terminals at the switching instant. Thus, for the worst case scenario where are 2 pu voltage at the CB terminals at the switching instant, it is possible for the voltage in the cable to reach a theoretical peak voltage of 3 pu if no damping is present.

When restrike/reignition occurs, there is also high-frequency component that is superimposed to the system frequency. The frequency of the high-frequency current is higher than the frequency of the power current and the current through the CB crosses zero. If the CB has the ability to extinguish this high-frequency current, the CB may open immediately after the restrike. If the CB is unable to interrupt the high-frequency current, it will only be able to open in the zero crossing of the power frequency.

This interruption of the high-frequency current can in some situations worsen the phenomenon and result in multiple restrikes, where the voltage increases until there is a cable failure or an external flashover.

Figure 4.27 shows an example of possible multiple restrikes in the CB of a capacitor.[29] It is given an order for the opening of the CB, which opens when the current crosses zero at t_1. The voltage in the capacitor remains equal to -1 pu after the disconnection while the voltage on the source continues to oscillate. At t_2 there is a restrike and a re-energisation of the capacitor. The voltage difference at the CB

---

[29] It is used a capacitor instead of a cable for simplicity. In the cable there would be reflections in the cable because of travelling waves and the voltage would be lower because of the resistance of the cable. This case is also the worst-case scenario where the restrikes occur at the worst instants.

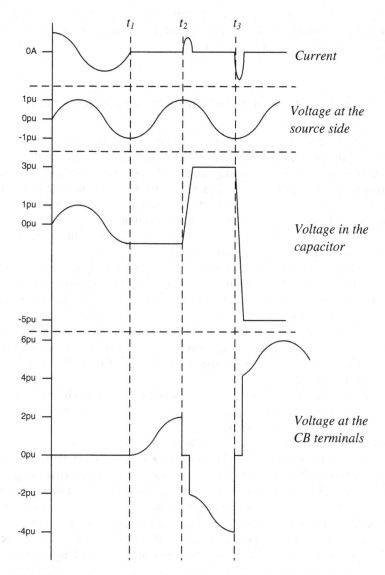

Fig. 4.27   Voltage and current during the multiple restrike of a capacitive load

terminals at this instant is of 2 pu, whereas the voltage in the capacitor is of −1 pu. Consequently, the voltage in the capacitor becomes 3 pu (−1 + 2 × 2). During the restrike there is also a current flowing in CB. The magnitude of the current and frequency is calculated as for a normal switching transient.

At t_3 there is another restrike. The waveforms are alike the ones of the previous restrike, but larger because the voltage at the terminals of the CB at the restrike instant is also larger.

It is important to refer that a restrike/reignition is just a disruption of the air in the CB and not a reclose of the contacts of the CB, which continue to separate from each other during the phenomenon.

4.7.1 Example and Influence of the Bonding Configuration

We have seen in Sect. 4.3 that the bonding configuration of the cable influences the magnitude of the peak voltage, because of the modal waves created at the points where the screens are crossed. The magnitude of the peak voltage in a restrike may be even larger because of the larger difference at the CB at the restrike instant for the worst-case scenario. Consequently, it is expected that the influence of the bonding configuration and the number of major-sections is larger as well as the difference between the peak values. The verification is made simulating a restrike in the 165 kV−50 km cable previously described.

The simulation consists in the CB being forced to re-close approximately half a cycle after the opening of the first phase (Phase A) so that the worst case can be simulated. The moment corresponding to the maximum overvoltage depends on several aspects, e.g., the cable parameters, short-circuit level, length, etc., making it very difficult to calculate it a priori. Therefore, a statistical switching (50 simulations) is run, using a Gaussian distribution with a standard deviation of 1.8 ms. Table 4.6 shows the maximum magnitude of the peak voltage in Phases A and C for the simulated restrike.

Contrary to expectations, it seems that the worst-case scenario is not to reclose the CB exactly half-cycle after opening of the first phase as the peak voltage is larger for Phase C than for Phase A.

In reality, the worst case is if the CB recloses some instants before the half-cycle point, the precise instant is difficult to estimate with high precision, thus, the use of statistical switching. The total voltage (V_T) is the sum of two components, the power frequency (V_P) and the high-frequency component (V_{HF}), being the peak overvoltage equal to the peak high-frequency voltage plus the power frequency in that instant

The peak voltage does not occur precisely after half-cycle, but hundreds of microseconds later, when the power voltage in Phase A is no longer at its peak value. The power frequency voltage (V_P) decreases in Phase A, while increasing in

Table 4.6 Maximum voltage at the cable receiving end after a restrike for different bonding configurations

	Peak voltage (Phase A)	Peak voltage (Phase C)
Both-ends bonding	2.47	2.63
Cross-bonding: 1 major-section	2.79	3.13
Cross-bonding: 2 major-sections	2.59	3.29
Cross-bonding: 3 major-sections	2.97	3.46
Cross-bonding: 4 major-sections	2.98	3.41

Fig. 4.28 Voltage at the receiving end of the cable due to a restrike. *Dotted line* both-ends bonding, *solid line* cross-bonded cable with 1 major-section, *dashed line* cross-bonded cable with 4 major-sections

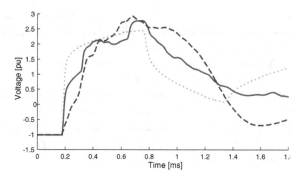

Phase C, between the CB reclosing and the peak overvoltage instants, following the normal 50 Hz sinusoidal curves. Therefore, it is possible to have a higher peak overvoltage in phase C than in Phase A. It should be noted, however, that even not having the higher overvoltage in Phase A, this phase continues to have a larger high-frequency component (V_{HF}).

Figure 4.28 shows the voltage at the receiving end of the cable due to a restrike for different bonding configurations. The larger number of small reflections in the cross-bonded cable is easily observed and in some situations they even reduce the voltage. The peak voltage values are different form the ones of Table 4.6 because in the simulation is simulated a restrike precisely 10 ms after the switch-off for all three bonding configurations.

4.7.2 Cable and Shunt Reactor

In Sect. 4.6 we studied the de-energisation of a cable connected to a shunt reactor and we saw that the voltage is no longer a decaying DC, but a decaying AC oscillating at resonance frequency.

Thus, the maximum voltage difference at the CB terminals is no longer half a cycle after the disconnection. For a lossless 50 Hz system, the time necessary to obtain a 2 pu voltage difference at the CB terminals is given by (4.41), which for a typical resonance frequency of 30–45 Hz it is equivalent to 2.5–10 cycles at power frequency.

If the resonance frequency is 50 Hz, the waveforms in both sides of the CB are in phase and initial with the same magnitude. The voltage gradient slowly increases with the damping of the voltage in the cable.

$$t = \left| \frac{1800}{180 - 3.6 f_r} \right|, \text{ t in milliseconds} \tag{4.41}$$

In the event of a restrike, if it occurs when the voltage at the CB terminals is maximum (i.e., 2 pu), the overvoltage has a magnitude similar to the one obtained when there is not a shunt reactor connected to the cable. However, as the maximum voltage difference at the CB terminals occurs later in time, the restrike is less likely to happen, as the TRV is less sharp

A final aspect that should be referred is the influence of the mutual coupling between the phases of the shunt reactor. We have seen that the coupling may increase the voltage on the cable on the first instant after the CB switch-off. As a result, the voltage at the CB terminals may in some situations be larger than 2 pu. However, the TRV continues to be less sharp and the maximum voltage at the CB terminals is attained after a relatively long period, substantially reducing the likelihood of the restrike.

4.8 Hybrid Cable-OHL

The use of hybrid cable-OHL lines[30] is a common occurrence when parts of a long OHL have to be replaced by cables for the crossing of long water courses, densely populated areas or spots of exceptional beauty.

The surge impedance of an OHL and a cable are different, resulting in reflections and refractions at the junction point between the cable and the OHL that change the behaviour of line when compared with a line with only a cable.

4.8.1 Energisation and Restrikes

To obtain a first idea of the phenomenon, we simulate the restrike for nine different cases, registering the maximum voltage magnitude associated to each case. We have seen that the theoretical explanations of the waveforms associated to a restrike are identical to those given for a normal energisation. However, it is easier to see the specificities of a hybrid cable-OHL in a restrike because of the large voltage difference at the terminals of the CB at the "switching" instant.

The simulations are made for the works case scenario, i.e., maximum voltage difference at the CB terminals at the re-switching instant, and using statistical switching:

- Case 1: Half cable-Half OHL
- Case 2: One third cable-Two thirds OHL
- Case 3: Two thirds cable-One third OHL
- Case 4: Half OHL-Half cable
- Case 5: Two thirds OHL-One third cable
- Case 6: One third OHL-Two thirds cable
- Case 7: One third Cable-One Third OHL-One Third Cable.

---

[30] The term syphon line is also commonly used.

- Case 8: One third OHL-One Third Cable-One Third OHL
- Case 9: Pure OHL

We have previously seen how the bonding of the cable may influence the waveforms and the magnitude of the peak voltage. Thus, the simulations are made for three different bonding configurations: both-ends bonding, one cross-bonded major-section and three cross-bonded major-sections.

Table 4.7 and Fig. 4.29 show the maximum peak voltage for the different scenarios.

The simulations show that the peak voltage is strongly influenced by the bonding configuration and the layout of the hybrid line. The former was explained

Table 4.7 Maximum voltage in the hybrid line receiving end during the re-energisation (*MS* major-section)

	Both-ends bonding		Cross-bonding: 1 MS		Cross-bonding: 3 MS	
	Max. Ph. A	Max. Ph. C	Max. Ph. A	Max. Ph. C	Max. Ph. A	Max. Ph. C
Pure Cable	2.47	2.63	2.79	3.13	2.97	3.45
Case 1	5.26	5.37	4.99	4.11	5.19	6.25
Case 2	5.59	5.74	6.13	6.48	6.42	7.16
Case 3	4.85	4.95	5.12	5.27	4.58	4.46
Case 4	2.59	2.87	2.63	2.90	2.64	2.95
Case 5	2.65	2.95	2.64	2.94	2.65	2.95
Case 6	2.58	2.85	2.63	2.98	2.64	2.96
Case 7	2.75	3.04	2.74	3.06	2.80	3.24
Case 8	2.67	2.96	2.85	3.19	2.94	3.49
Case 9	3.00	3.00	3.00	3.15	3.15	3.15

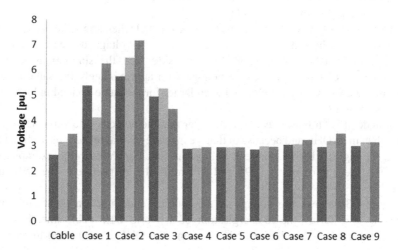

Fig. 4.29 Maximum voltage in the hybrid line receiving end during the re-energisation. *Left column* both-ends bonding, *middle column* one cross-bonded major-section, *right column* three cross-bonded major-sections

in previous sections, whereas the latter is a result of different travelling times and reflections/refractions at the junction point(s) between the cable(s) and the OHL(s).

The reflected and refracted voltages are calculated for two general surge impedances by (4.42), where V_1 is the sending voltage, V_2 is the reflected voltage, V_3 the refracted voltage, Z_A and Z_B the surge impedance of the lines.[31]

$$V_3 = V_1 \frac{2Z_B}{Z_A + Z_B}$$
$$V_2 = V_1 \frac{Z_B - Z_A}{Z_A + Z_B}$$
(4.42)

The surge impedance of an OHL is typically higher than the surge impedance of a cable, resulting in a voltage reduction when the wave flows from an OHL into a cable and a voltage magnification when the wave flows from a cable into an OHL.

By substituting in (4.42) the cable and OHL surge impedances, respectively, 38.8 and 462.6 Ω for the simulations previously done, (4.43) and (4.44) are obtained for an incident wave flowing from the cable into the OHL and (4.45) and (4.46) for an incident wave flowing from the OHL into the cable.

$$V_3 = V_1 \cdot 1.845$$
(4.43)

$$V_2 = V_1 \cdot 0.845$$
(4.44)

$$V_3 = V_1 \cdot 0.155$$
(4.45)

$$V_2 = V_1 \cdot (-0.845)$$
(4.46)

As it can be seen in (4.43), the voltage should ideally increase 1.845 times when the restrike occurs in the cable end, explaining the voltage increase when comparing cases 1–3 with a pure cable line.

In the same way that a voltage increase is expected when a restrike occurs in the cable side of a hybrid line, according to (4.45) a voltage decrease would be expected if the restrike occurs at the OHL side. Yet, the simulations seem to contradict the equation and show a voltage with approximately the same magnitude for the cross-bonded cables and even larger for the cable bonded in both-ends (cases 4, 5 and 6).

To understand this, we need to remember that the velocity of a wave in cable is lower than in an OHL and because of that we may see several reflections reaching the line receiving end before the peak voltage instant. For this specific example, the velocities are approximately 298 and 176 m/µs for, respectively, the OHL and the cable.

Figure 4.30 shows the voltage at the sending end, junction point and cable receiving end during the re-energisation of a hybrid OHL-cable line. For simplicity, the cable is bonded in both-ends, and the OHL is connected to an ideal

---

[31] An introduction to the topic was made in Sect. 3.6.

Fig. 4.30 Voltage at the sending end, junction point and cable receiving end during re-energisation of OHL-cable line. *Dotted line* sending end, *dashed line* junction point, *solid line* receiving end

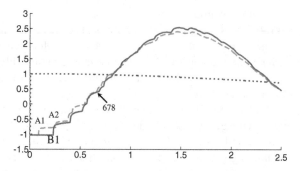

voltage source, resulting in a reflection coefficient of -1 at the OHL sending end. For a cross-bonded cable, there would be even more reflections/refractions as we have seen in previous sections.

The voltage difference at the CB terminals at the restrike instant is 2.028 pu[32] and the wave travelling times are 83 and 143 μs, for, respectively, the OHL and the cable.

At $A1$ (83 μs), the voltage wave reaches the junction point between the OHL and the cable. In a lossless line approximately 0.31 pu (2×0.155) would propagate into the cable, while -1.69 pu ($2 \times (-0.845)$) would reflect back into the OHL. Figure 4.30 shows a voltage in the junction point of -0.79 pu $\sim(-1 + 0.31)$ at $A1$, a value that can still be considered as being inside of the expected interval.

At $B1$ (226 μs), the wave reaches the receiving end of the line. We know from $A1$ that approximately 0.21 pu were injected into the cable. The cable is open and this value is to be reflected back in the receiving end, resulting in an expected voltage of -0.56 pu ($-1 + 2 \times 0.21$). The value obtained in the simulation was -0.64, being the difference because of the damping of the wave in the cable.

At $A2$ the waves that reflected back into $A1$ reaches the junction after being reflected at the sending end. The behaviour at $A2$ is alike to $A1$, but with a smaller gradient because of the lower magnitude of the wave. It is important to remember that we have a double negative in this situation. At $A1$, the polarity of the wave reflected back into the OHL was reversed and the wave polarity was reversed again by the ideal voltage source at the sending end.

We can continue to use the same procedure to explain the remaining variations of the waveforms in what would be a tedious and long process. Instead, we focus our attention in what would be the expected peak voltage instant.

We have previously seen that for a cable connected to an ideal voltage source the peak voltage is achieved when the wave reaches the cable receiving end for the second time, i.e., after being reflected at the cable receiving end and at the source.

---

[32] The voltage of an OHL increases after the disconnection because of the capacitance coupling between the phases. Consequently, the voltage at the CB terminal at the restrike instant is larger than 2 pu. However, we consider the value 2 pu in the demonstration of the phenomenon.

In this example that would correspond to a time of 678 μs (83 + 2 × 143 + 2 × 83 + 143). Instead the peak voltage is at approximately 1,480 μs and it is not as easy to detect as for a cable alone. Moreover, we can see that while for a cable alone the voltage at the receiving end varies only two times before the peak instant (see Fig. 4.2) for this hybrid line with the same length there are more than a dozen variations.

We have to remember that the reflection coefficient in the junction point completely changes the magnitude of the wave, creating also several other waves. As an example and assuming lossless lines, the peak voltage at the receiving end of a cable would be given by (4.47), where the first two terms are the first incident wave and respective reflection and the last two are the wave after being reflected at the sending with opposite polarity. The same voltage for the hybrid line is given by (4.48).

$$V_P = [2 + 2] + [-2 + (-2)] = 0 \tag{4.47}$$

$$V_P = [(2 \times 0.155) \times 2] + [(2 \times 0.155 \times 1.845 \times (-1) \times 0.155) \times 2] = 0.44 \tag{4.48}$$

According to (4.48) there would be a drop in the magnitude of the voltage at 678 μs, but the simulation of Fig. 4.30 shows a voltage increase at that instant. To understand that apparently contradictory result lets follow the wave reflected back into the OHL at $A1$ (83 μs). That wave goes back to the source and is reflected back into the OHL reaching the junction point at $A2$ (83 μs + 2 × 83 μs). Part of this wave is refracted into the cable reaching the cable receiving end 143 μs later when it is fully reflected back. After another 143 μs, the wave reaches the junction point and part of the wave is reflected back into the cable. This reflection reaches the cable receiving end after another 143 μs. If we add all these travelling times, we obtain a final value of 678 μs and we see that this wave reaches the receiving end of the line at the same time than the wave that is expected to reduce the magnitude of the voltage. The peak voltage at the receiving end associated to this wave is given by (4.49).

The change in the magnitude of the voltage at 678 μs is given in the second part of the equation, which has a positive signal. If we add the second parts of (4.48) (−0.1773 pu) and (4.49) (+0.4427 pu), we obtain a final result of 0.2654 pu, explaining the positive variation of the voltage at 678 μs.

$$\begin{aligned} V_P =& [(2 \times (-0.845) \times (-1) \times 0.155) \times 2] \\ &+ [((2 \times (-0.845) \times (-1) \times 0.155) \times 0.845) \times 2] \\ =& 0.966 \end{aligned} \tag{4.49}$$

These calculations show that just because of the reflections at the junction point the estimation of the peak voltage instant is much more complex. As a matter of fact, for a hybrid line we can no longer previously predict when the peak voltage occurs, unless the cable is much larger than the OHL or vice versa.

Even the magnitude of the peak voltage is seriously affected by the lengths of the cable and/or the OHL and can have a large variation just by an increase or decrease of some hundreds of meters in the length of the cable or the OHL.

4.8.2 Summary

The energisation of hybrid lines was explained in this section and it was demonstrated how the difference between the surge impedance of a cable and an OHL may change the waveforms, the magnitude of the peak voltages and the respective instant of occurrence.

The worst case scenario is to have the wave flowing from a cable to an OHL as there is a magnification of the voltage at the junction point. As an example, we saw in the simulations an increase of more than 100 % in the magnitude of the peak voltage for the one third cable-two thirds OHL case when compared with a cable of equal length. Different configurations and/or cable/OHL geometries may result in even larger differences.

For a wave flowing from the OHL to a cable there is a large reduction of the voltage, but because of all the reflections/refractions and the waves generated at the junction point, the peak voltage can still be larger than for a cable of equal length.

We have also seen how difficult it is to estimate the instant at which there is maximum voltage and how it is influenced by the travelling times and the reflections/refractions coefficients.

In theory there is a similar phenomenon when two cables with different characteristics are connected to each other.[33] However, the difference between the two surge impedances is normally so low that the phenomenon is not noticeable.

Finally, a word for the influence of the bonding configurations, which is much more relevant in a hybrid line, because of the larger voltages associated to the phenomenon.

4.9 Interaction Between Cables and Transformers

4.9.1 Series Resonance

The series inductance of a transformer can create a resonance circuit with the capacitance of a cable, influencing the transient waveforms.

---

[33] A good example of this type of line is the connection of an offshore wind farm, where part of the line is a submarine cable and the rest a land cable.

We start by analysing the simplest case: a transformer in series with a cable, which are energised together by an ideal voltage source, as shown in Fig. 4.31.

Tables 4.8 and 4.9 show the frequency and the impedance of the series resonance point seen from the transformer primary for two different transformer's leakages reactances and six cable lengths.[34] The frequency of the resonance point decreases as the length of the cable increases, which is explained by increase of the capacitance of the cable. The same happens when the leakage reactance of the transformer increases, as the inductance of the transformer increases. The magnitude of the series resonance has also a discernible behaviour and it increases as the capacitance of the cable increases and the inductance of the transformer increases.

The cable is energised through the transformer and because of the high inductance of the last there is no longer present the switching overvoltage described in Sect. 4.3. Instead the waveform can be seen as the summation of two sinusoidal components, a steady-state component and a transient component. The frequency of the transient component is the frequency of the series resonance and the damping is proportional to the negative exponential of the magnitude of the impedance at the resonance frequency point.

From a theoretical point of view, the phenomenon can be seen as alike to the energisation of a RLC load. Actually, if we design an equivalent RLC circuit and simulate its energisation, we will obtain waveforms that are very similar to the ones obtained using a transformer model and a cable model.

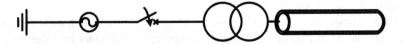

Fig. 4.31 Example of a transformer-cable circuit

Table 4.8 Frequency and magnitude of the series resonance point for a transformer with a leakage reactance of 0.1 pu

	10 km	20 km	30 km	40 km	50 km	60 km
Frequency (Hz)	346	243	198	170	152	138
Magnitude (Ω)	13.1	15.8	18.2	20.2	22.2	23.8

Table 4.9 Frequency and magnitude of the series resonance point for a transformer with a leakage reactance of 0.0335 pu

	10 km	20 km	30 km	40 km	50 km	60 km
Frequency (Hz)	593	414	334	286	252	228
Magnitude (Ω)	15.7	19.6	22.8	25.6	28.1	30.3

---

[34] The bonding of the cable influences the frequency spectrum as demonstrated in Sect. 3.5. For simplicity, in this section the cable is considered as being bonded in both-ends.

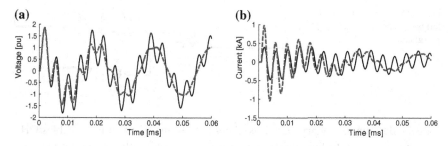

Fig. 4.32 Waveforms during the energisation of a cable through a transformer. **a** Voltage. **b** Current. *Solid line* 20 km cable energised by a transformer with a leakage reactance of 0.1 pu, *dashed line* 50 km cable energised by a transformer with a leakage reactance of 0.0335 pu

Figure 4.32 shows the voltage and the current during the energisation of a transformer-cable system. The first simulation system is the 20 km cable connected to a transformer with a leakage reactance of 0.1 pu. According to Table 4.8, the resonance frequency of this system is 243 Hz and the impedance at the resonance frequency is 15.8 Ω. The second system is the 50 km cable connected to the 0.0355 pu, which has a resonance frequency of 252 Hz and an impedance of 28.1 Ω at that frequency. The simulations confirm the resonance frequency of the two transient components and the respective dampings.

The simulations also show that the duration of the transient is much longer than when the cable is connected directly to a voltage source, an undesired situation. This is explained by the low impedance of the system at the resonance frequency, which is also the frequency of the transient component. When the cable is energised directly by a voltage source there is a high-frequency component, whose associated impedance is high resulting in a faster damping.

As it was stated the phenomenon is in these conditions identical to the energisation of a RLC circuit, which as we have seen in Chap. 2 depends on the switching instant. In the examples shown in Fig. 4.32, the energisation is made for peak voltage. If the energisation was made for zero voltage, a transient would still be present, but the transient component would be much lower.

Meshed Grid

We have seen how the phenomenon processes for a system with only a voltage source, a transformer, a cable and a CB. However, a simplistic system like this one is very unlikely in real life.

It is very unlikely to energise the cable and the transformer together. For the simplest system used in the previous example that would be irrelevant and the transients would be identical for either the energisation of the transformer and the cable together or for the energisation of the cable after energising the transformer, but the same does not happen in a real system.

It is usual to have several lines connected to a busbar. For the moment, we consider the transformer to have a 400/150 kV transformation ratio and the cable

Fig. 4.33 Single-line for a series resonance due to cable energisation

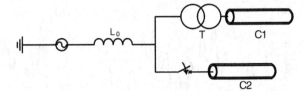

connected to the secondary side. It is likely in this situation that more cables/OHL are connected also to the busbar and already energised at the switching instant.

Thus, the cable is no longer fully energised through the transformer, instead only part of the energy is provided through the transformer, whereas the rest comes from the line(s) connected to the busbar. The amount of power being provided through the transformer depends on several factors: the length of the line(s) connected to the busbar, the short-circuit power level, the inductance of the transformer. As an example, if the lines are connected to weak networks in the other end or are very long the majority of the power during the transient continues to be provided through the transformer. By other hand, if the lines are short and connected to a strong network less power is provided through the transformer and the transient is similar to the switching transient explained in Sect. 4.3.

Energisation of cables in parallel
Another situation when a series resonance may occur is when energising a cable in parallel to a transformer + cable system already energised.

Figure 4.33 illustrates the phenomenon. The inductance of the transformer creates a series resonance circuit with the capacitance of the cable (*C1*) for a given frequency. The energisation of the second cable (*C2*) generates inrush currents whose frequency depends on the length and electrical parameters of the cable. The impedance of the transformer + cable circuit at the series resonance point is very low, thus, a current with the same frequency flows mostly to the resonance circuit instead of the source.

Therefore, if the frequency of the inrush current matches the frequency of the series resonance circuit, an overvoltage appears on the secondary side of the transformer as most of the current will flow into the transformer + circuit circuit.

4.9.2 Parallel Resonance

A parallel resonance is characterised by a large impedance at the resonant frequency, which may generate high voltages if excited by a current of equal frequency.

A situation where this may happen is the energisation of a transformer through a long cable in a weak network, as shown in Fig. 4.34. The shunt capacitance of the cable can be seen as being in parallel with the inductance of the equivalent network[35] and of the shunt reactor, if present. The frequency of the parallel resonance is inversely proportional to the square root of the capacitance of the cable, the inductance of equivalent network and shunt reactor. Consequently, the frequency is lower for long cables and weak networks.

Depending on the switching angle, the energisation of a transformer may generate inrush currents containing all the harmonics, if one matches the frequency of the parallel resonance of the circuit an overvoltage is present.

A transformer is an inductive element, which as we know draw larger current when energised at low voltages.[36] When the current in a transformer exceeds the saturation current, the impedance of the transformer is strongly reduced as the inductance drops to very low values. Consequently, the magnitude of the resulting current increases to very high values.

The current in the transformer is oscillatory and there is saturation for only approximately half-cycle. As a result, harmonic currents are generated and propagated into the system. In some situations, this is also called a parallel ferroresonance.

Typically, the larger the transformer, the longer it is the duration of the inrush current, because of the larger X/R ratio. Other factors influencing the inrush current magnitude and duration are the residual flux and the magnetisation characteristic.

Example
In the example is used the same cable of the previous examples with a length of 100 km, compensated at 70 % and connected to a network with a short-circuit power of 870 MVA. The parallel resonance frequency of this network is at approximately 100 Hz, which corresponds to the second harmonic.

After the cable reached steady-state conditions, a transformer connected to the cable receiving end with a switching angle of 0° (zero volts) is energised. The energisation of the transformer generates an inrush current containing all the harmonics, including the second, which excites the parallel resonance circuit and generate an overvoltage.

Figure 4.35 shows the voltage and the current in the transformer during its energisation and it is observed a transient overvoltage (TOV) with the duration of

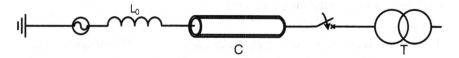

Fig. 4.34 Single-line for a parallel resonance due to transformer energisation

---

[35] The inductance of the equivalent network is an artificial representation of the network for effects of representation, but the physical behaviour is similar.

[36] From a more formal point of view, the energisation at zero voltage results in maximum flux that corresponds to a near saturation point in the B–H curve.

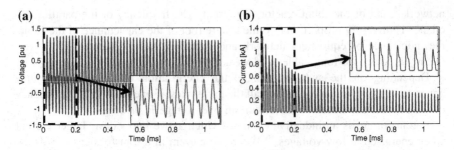

Fig. 4.35 Voltage and current during the energisation of the transformer. **a** Voltage in the transformer. **b** Current in the transformer primary

several cycles. The transformer's inrush current is also observed during the first cycles of the energisation.

As previously stated, the phenomenon depends on the frequency of the parallel resonance. As an example, if the cable length was 80 km the resonance frequency would be approximately 120 Hz. The current in the transformer would still be equal to the one shown in Fig. 4.35b, but the magnitude and duration of the overvoltage would be smaller, since there is not a high impedance at 100 Hz anymore.

4.9.3 Ferroresonance

Ferroresonance is one of the most complex and challenging phenomenon presented in this book. This phenomenon was observed for the first time more than 100 years ago and as being known for more than 90 years. In a nutshell, ferroresonance is defined as an interaction between capacitors and saturable iron-core inductors, thus, the terms *resonance* and *ferro*, respectively.

From a practical point of view, it can occur when a transformer and a cable become isolated, and the cable's capacitance is in series with the transformer magnetising characteristic, which may lead to sustained temporary overvoltages of long duration that are the most undesired.

Theoretical Description
We start by considering a simple circuit consisting in an AC voltage source, an inductor and a capacitor all connected in series. We already know that for a circuit like this, the reactance of the inductor (X_L) cancels the reactance of the capacitor (X_C) for a given frequency and there is a series resonance.

However, a transformer is not properly represented by a linear inductance, because of its saturation. Thus, if we want to have a more accurate model, we have to replace the linear inductor by a non-linear inductor.

Figure 4.36 shows the voltage in function of the current, where V_S is the voltage of the source, V_L is the voltage of the inductor and V_C is the voltage of the capacitor and $C > C2$. The solution of the circuit is given by (4.50).

$$V_S = V_L + V_C \tag{4.50}$$

Figure 4.36 shows that there are three possible operation points:

- Point 1: It is the non-ferroresonance stable point, i.e., it is the point that would also be obtained with a linear inductance. It corresponds to the inductive solution of the circuit $(X_{L\_LIN} > X_C)$;
- Point 2: It is the ferroresonance stable-point, with both large currents and voltages. It corresponds to the capacitive solution of the circuit $(X_C > X_{L\_SAT})$;
- Point 3: It is an unstable point and the solution will not remain there for steady-state conditions;

Thus, a ferroresonance circuit can have two stable operation points and there may be sudden variation of voltage and current from one stable point to other.

To understand this behaviour, we analyse a LC circuit with a saturable inductor connected to an AC voltage source. The circuit is energised and a current and a voltage oscillating at the resonance frequency $(1/\sqrt{L_{LIN}C})$ appear, as explained in Chap. 2. This corresponds to the operation point 1.

As the system oscillates, the voltage and the current increase and at some point the inductor saturates and the current increases for very high values. Another way of seeing the phenomenon is to consider that the inductance of the saturable inductor decreases to a value dozens of times smaller, i.e., $L_{SAT} \ll L$. Consequently,

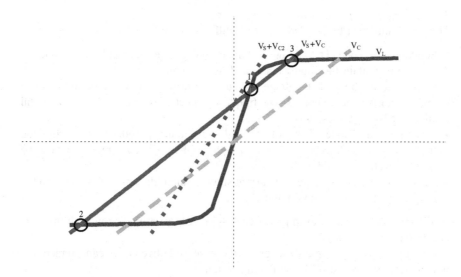

Fig. 4.36 Graphical solution of a ferroresonance circuit

$X_C > X_{L\_SAT}$ and the behaviour changes as well as the resonance frequency, which is now given by $1/\sqrt{L_{SAT}C}$, whereas the system operation point is point 2.

Moreover, as the inductance value is now very small the current reaches very high value, whose peak value is achieved when the magnetic energy in the inductor is equal to the electrical energy just prior to the situation, as given by (4.51), where V is the voltage at the capacitor terminals at the saturation instant.

$$\frac{1}{2}L_{SAT}I^2 = \frac{1}{2}CV^2 \Leftrightarrow I = V\sqrt{\frac{C}{L_{SAT}}} \tag{4.51}$$

Half-cycle after the saturation, at the resonance frequency of the saturated circuit, the flux is below the saturation limit and the inductance of the saturable inductor is again L_{LIN}, i.e., the system is back to the operation point 1. During this time, the voltage changed polarity and has a magnitude at the "unsaturation" instant that is approximately the same that it had just before the saturation but with opposite signal.

In this operation conditions, the system will continue to oscillate between this two operations point ad aeternum.

Thus, if we think about it, what is happening here is almost the same that happens in a normal LC circuit, but with two values for the inductance, which result in sudden changes in the waveforms that make them seem unpredictable and larger magnitudes because of the low inductance of the inductor when saturated.

Figure 4.37 shows an example of ferroresonance. The horizontal dotted lines represent the saturation limits and the system saturates when the current (solid line) crosses them. When this happens, there is a fast increase of the current's magnitude and a fast variation of the voltage's magnitude that has approximately a symmetrical magnitude when the system goes back to the solution point 1.

Ferroresonance types and initial conditions
Ferroresonance can be divided into four different types:

- Fundamental mode: The voltage and the current are periodic and their frequency spectrums contain the fundamental and its harmonics (nf).
- Sub-harmonic mode: The voltage and current are periodic with a period that is a multiple of the source period. The frequency spectrums contain the fundamental and its sub-harmonics (f/n).
- Quasi-periodic mode: The voltage and current are pseudo-periodic and the frequencies are defined in the form $nf_1 + mf_2$, where n and m are integers and f_1/f_2 is an irrational number.
- Chaotic mode: The voltage and current show an unpredictable behaviour and the frequency spectrum is continuous.

Figure 4.38 shows an example of each type of ferroresonance in both the time and frequency domains.

The ferroresonance type depends on the relation between the capacitance and inductance, and on the initial switching instant.

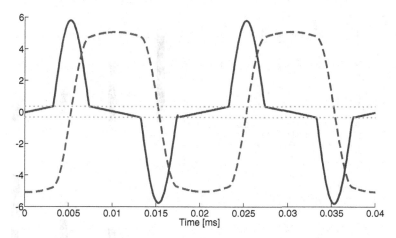

Fig. 4.37   Example of ferroresonance in RLC circuit (not at scale). *Solid line* current, *dashed line* voltage at the terminals of the capacitor

The most common types are the two periodic types: *fundamental mode* and *subharmonic mode*. These two modes can only occur if (4.52) is true. For capacitance values outside this interval it is still possible to have ferroresonance, but its type will be either *quasi-periodic mode* or *chaotic mode*.

$$\frac{\omega_0 L_{\mathrm{LIN}}}{n} < \frac{n}{\omega_0 C} < \frac{\omega_0 L_{\mathrm{SAT}}}{n} \tag{4.52}$$

We have not included the losses in our analysis, but as expected the probability of having ferroresonance decreases when the losses increase.

The appearance of ferroresonance is more likely if the system is energised at zero volts, because of the higher transient currents associated to this switching angle when energising an inductive load.[37] As the current is larger, the saturation is also more likely and consequently the same goes for the appearance of ferroresonance.

Similarly, the existence of remanent flux in the saturable inductor increases also the likelihood of having ferroresonance, as saturation is more probable.

Finally, the magnitude of the voltage also influences the phenomenon as the magnitude of the current is proportional to it.

Cable-transformer system

Having seen how the ferroresonance may occur in an ideal LC circuit, it is time to study more realistic systems.

One of the conditions necessary for the appearance of the phenomenon is the system to be lightly loaded or to have a low short-circuit power. The phenomenon

---

[37] The same happens for the energisation of a transformer, a phenomenon known as inrush currents.

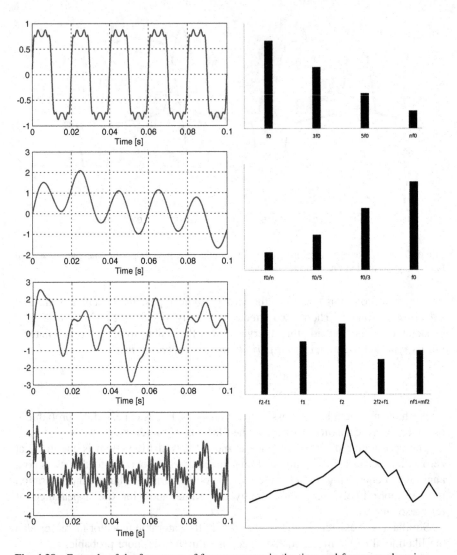

Fig. 4.38 Example of the four types of ferroresonance in the time and frequency domains

is also worse for long cables, because of the larger capacitance. The operation point 2 for a system with a low capacitance, i.e., a short cable, does not result in high voltages.

There are several configurations that may lead to the occurrence of the phenomenon, especially when using instrumentation transformers, which are typically lightly loaded and have low losses. We will not analyse those cases, instead we will concentrate our attention in system where the cable is connected to a power transformer.

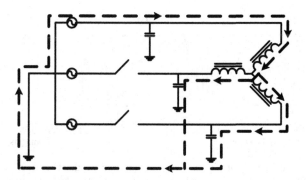

Fig. 4.39 Possible ferroresonance scenario in a cable-transformer system

It is very unlikely to have ferroresonance in a cable-transformer that is operating normally. The exception is for very weak networks.

However, the cable capacitance and the transformer inductance may be in series if there are problems with the poles of the CB and they are on different switching states. i.e., some of the pole(s) open while the remaining stay close, for a long period of time.

Figure 4.39 shows an example of the phenomenon for the energisation of a cable (represented by a capacitor to ground) and a transformer, where only one of the phases closes and the transformer has a star configuration without a grounded neutral. The capacitance of the cable at the two open phases is in series with the inductance of the transformer, as indicated by the arrows, creating a possible ferroresonance circuit.

The same situation may occur if two of the phases close while one remains open or during a de-energisation if not all the phases open.

The type of connection used for the transformer influences the appearance of the phenomenon. The analysis would be the same if the transformer was connected in delta formation. However, if the transformer neutral was grounded, ferroresonance would be very unlikely as the majority of the return current would flow to ground through the neutral of the transformer and not the "capacitor(s)".

Figure 4.40 shows an example of ferroresonance for the disconnection of a cable-transformer system where one of the phases does not open. The simulations were made for a Δ/Yn and Yn/Yn transformer connections, and it is possible to observe that there is overvoltage only for the former.

It is important to remember that to have a system operating as described is a necessary condition for ferroresonance, but it is not sufficient. The phenomenon depends also on the residual flux in the transformer, the switching instant and the values of the capacitance and inductance.

Modelling

A proper modelling of all the equipment is required for a proper simulation of ferroresonance. The phenomenon is non-linear, meaning that we cannot use

(a)   **(b)**

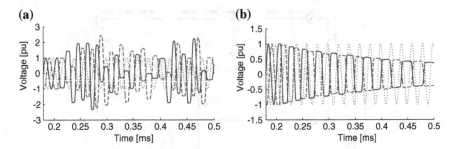

Fig. 4.40 Voltage in the cable after a faulted opening of the CB at 0.2 s. **a** Delta-Star connection. **b** Star–Star with both neutrals grounded. *Solid* and *dashed lines* open phases, *dotted line* close phase

lumped parameters. It is also frequency dependent, making it necessary to use FD-models.

The modelling of the cable continues to be made like done in the previous sections and chapters of the book using the FD-models already explained.

The modelling of the transformer is more challenging, as there are several transformer types, e.g., three-legs, four legs, core or shell type, etc., and several modelling possibilities.

The most important is to use the correct saturation data and design an appropriated voltage-current curve or a flux-current curve. This data is normally available on the transformer's datasheet, but need to be adapted for some software.

A proper modelling of the series losses is not so important. The voltage at the transformer's coil is higher than the rated voltage during the phenomenon. Consequently, the shunt losses are much larger than the series losses. For the typical materials used in the construction of a transformer it is not necessary to consider hysteresis when studying ferroresonance.

The modelling for high frequencies is more complicated and it would be necessary to stray capacitances and the inner windings in the model. This information is difficult to obtain and results in a substantial increase of the model's complexity. However, this is not normally considered when studying the phenomenon, as the associated frequencies are not so high.

4.10 Faults

A great majority of transients in a power network are directly or indirectly associated to a fault. We already studied the transient recovery voltage (Sect. 4.7) and how the waveforms would be for the disconnection of a sound cable. The next step is to study how a fault influences the voltage and current waveforms and consequently also the transient recovery voltage.

4.10.1 Single-Phase

As usual we start with a single-phase cable circuit and a single-phase to ground fault, for an easier acquisition of the knowledge.

When a fault occurs, there is a decrease of the voltage at the fault location for virtually zero volts and an increase of the current as the impedance of the circuit is much reduced. The phase difference between the voltage and current is also affected by the fault.

Further, if we consider the cable as being and infinite number or pi-sections in series and a fault at a random point the voltage in that point goes to zero. As a result, the voltage between the sending end, which is equal to 1 pu when connected to an ideal voltage source, and the fault point that is 0 pu decreases steadily between these two points. As the voltage along the cable is lower, the voltage to the ground is also lower, meaning that the capacitive current is lower.

We learned that for a normal cable disconnection, the voltage in the cable would decrease very slowly. The opposite happens during a fault, when all the energy stored in the cable is discharged in some microseconds through the ground at the fault location.

Thus, it seems that the likelihood of having a reignition or a restrike is lower when the cable is disconnected because of a fault. However, we have to remember that this is a single-phase cable, for a three-phase cable there are some more aspects to consider, namely the coupling between phases. However, before ana-lysing these cases, we should study the waveforms for a single-phase case.

A fault may occur at any point of a cable, for just some meters to several kilometres from one of the extremes, influencing the shape of the TRV.

For simplicity, we start our analysis not for a cable, but for a single-phase 30 km OHL connected to an ideal voltage source. Single-phase to ground faults are simulated for different points of the OHL.

We saw in Sect. 4.6 that the voltage in a line during a de-energisation is a decaying DC waveform and the TRV is a sinusoidal wave oscillating at power frequency. However, in the simulations shown in Fig. 4.41 the TRV waveform has a saw-tooth shape on the first instants after the disconnection, especially for the faults closer to the CB, becoming more sinusoidal after some high-frequency cycles.

The saw-tooth waveform is a result of the reflections between the CB and the fault point. It is clear that at the switch-off instant the voltage at the fault location is zero, while having a higher value at the sending end, decreasing in approximately linear way between the two points. After the disconnection the charges in the OHL need to balance themselves resulting in travelling waves.

At this instant two waves are generated, one propagating in the forward direction of the line and other propagating in the backwards direction.[38] The waves

---

[38] The introduction to travelling waves and the explanation on why a wave is divided into two waves propagating in opposite directions is given in Chap. 3.

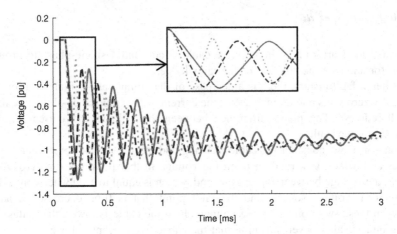

Fig. 4.41 TRV for different fault locations. **a** *Solid line* 15 km, *dashed line* 10 km, *dotted line* 5 km

are triangular along the line, in order to represent the voltage in the line prior to the disconnection. The magnitude of each wave along the line is equal to half of the voltage along the line prior to the disconnection. The total voltage wave is given by the summation of these two waves.

The waves are reflected between the CB and the fault until eventually disappear because of the losses. In this process, it is generated the saw-tooth shape seen in the simulations.

Figure 4.42 shows the forward and backward voltage waves along the line for different time instants in a lossless line. Figure 4.43 shows the voltage at the sending end and middle of the line in function of time. The figure shows that that the voltage at the sending end is a triangle, explaining the saw-tooth shape of the TRV.

In a real line with damping and inductance, the waves would lose the triangular shape after some instant and become more sinusoidal, explaining the reason for the saw-tooth waveforms in the simulations be only seen in the first instants of the disconnection and for faults closer to the CB.

Having understood the phenomenon for an OHL, we are ready to go to the next level and study it for a cable. In Chap. 1, there were described the typical layers of a cable and it was explained that one function of the screen is to provide a circulatory path for the fault currents. From here, it is directly deducted that there are differences in the waveforms associated to a short-circuit in a cable and one in an OHL, and that the bonding configuration is also influencing the waveforms.

If the cable was only the conductor and the insulation, the waveforms would be similar to the ones shown for the OHL, but with a much lower propagation speed.[39]

---

[39] In this case, the wave speed is much lower than the typical coaxial mode velocities because of the larger inductance of the cable when it does not have a screen. However, for practical reasons a HV cable without a screen would never be installed directly on the ground.

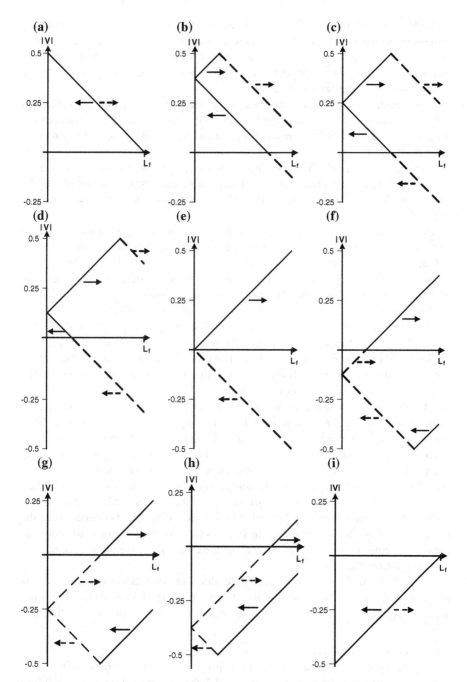

Fig. 4.42 Voltage waves in a faulted OHL after the switch-off of the CB. *Solid line* forward wave, *dashed line* backward wave. **a** t = 0. **b** t = 1/4v. **c** t = 1/2v. **d** t = 3 1/4v. **e** t = 1/v. **f** t = 5 1/4v. **g** t = 3 1/2v. **h** t = 7 1/4v. **i** t = 2 1/v

The conductor of a cable is enclosed by the screen, thus, it is virtually impossible to have an external fault to the conductor without having also a fault to the screen. Consequently, the voltage at the fault location is zero volts in both core and screen,[40] increasing approximately linearly in magnitude along the conductor as the distance to the fault increases.

It is true that the screen is grounded at the end of the cable and the voltage in that point would ideally be zero volts. However, the screen is connected to the ground through a grounding resistance, whose typical value is in the order of several Ohms.[41]

The exact value of the grounding resistance affects the phase difference between voltage and current during a fault. Consequently, the magnitude of the voltage at the core and screen of the cable at the disconnection instant is also affected.

During a fault, the conductor and the screen are short-circuited to the ground and the current increases to very high values. In these conditions, the system is unbalanced and a ground current exists. This return current will then divide itself through the ground and the screen, depending on the impedance values of each, which for the ground includes also the grounding resistance.

The observation of the two impedances shows that typically the screen is mostly resistive, whereas the earth is mostly inductive. The magnitude of the current flowing in the screen depends only on the distance between the fault and the end of the cable as well as the impedance of the screen, whereas the current flowing on the ground depends on the resistivity of the ground, the distance to the fault and the value of the grounding resistance.[42]

The magnitude of the current also increases when the grounding resistance value decreases, as there is more current flowing in the ground. Figure 4.44a shows the single-line equivalent circuit for a fault in a single-phase cable. The fault is to the ground, however, the potential at the fault point is not equal to the potential at the end of the cable and the latter is used as reference. The screen's impedance is connected to the grounding grid of the substation, which is then connected to the ground through a grounding resistance. Typically, a transformer with at least a star connection grounded to the grounding grid is connected to the cable, instead of the source shown in Fig. 4.44a. However, for now we consider the cable connected to a source and analyse the cases where the cable is connected to a transformer latter.

It is not so usual, but in more recondite areas, or when doing reparations, it is possible to find a configuration like the one shown in Fig. 4.44b, where there is not a grounding grid and the screen is connected to the ground. In this situation, the

---

[40] For simplification, we consider an ideal short-circuit with a fault resistance of 0 Ω.

[41] It is important not to confuse neutral with ground. If there is a grounding grid, this grounding grid is the neutral point, however, it will have a voltage slightly different form the ground.

[42] There are other factors influencing the current like the configuration of the substation, however, these parameters are typical of less importance.

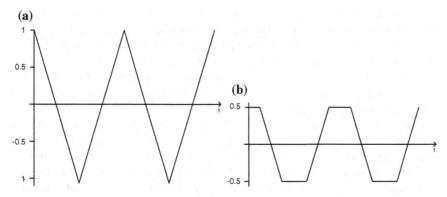

Fig. 4.43 Voltage in two points of the line. **a** Load side of the CB, **b** Middle of the line

grounding resistor influences the magnitude of the current in the screen instead of the magnitude of the current in the ground. A similar situation may happen in hybrid lines where the screen is grounded at the end of the cable section and the conductor of the cable connected to the conductor of the OHL.

In the next pages, we will analyse both scenarios side-by-side. It is true that the second scenario is less likely to happen, but its study will help us understanding some concepts related with the distribution of the current by the screens and the ground, for faults in three-phase cables with different bonding configurations. Moreover if the grounding resistance of Fig. 4.44b is very low, the results will be very similar to those obtained using the grounding grid, as we will see.

The high current associated to the short-circuit returns in the screen and in the ground. The latter is mostly inductive and so a transitory DC component appears on the current. The magnitude of the DC component depends on the magnitude of voltage in the conductor at the instant of the fault. If the voltage is zero, the component is maximum, if the voltage is at peak value there is not DC component.[43]

In the configuration shown in Fig. 4.44a, the current returning in the screen is independent of the grounding resistor, while the current returning in the ground decreases when the grounding resistor increases, resulting also in a reduction of the DC component. Thus, the higher the grounding resistance the lower is the DC component when compared with the AC component.

In the configuration shown in Fig. 4.44b, the DC component remains nearly constant and the AC increases when the grounding resistance decreases, being the differences in the DC component a result of a small change in the phase difference between voltage and current when the grounding resistance changes. Thus, the higher the grounding resistance, the larger is the DC component when compared with the AC component, a situation that is the complete opposite of the one shown

---

[43] If we think in the ground as being a small resistor in series with a large inductor, the phenomenon is exactly like the energisation of a shunt reactor.

to Fig. 4.44a. This DC component is damped after several cycles, but as the CB typically opens as fast as possible the DC component can still be large at the disconnection instant. In other words, in this scenario both the phase angle and the relative DC current increase when the grounding resistance increases.

Figure 4.45 shows an example of the results just described for a fault at the middle of a 5 km long cable for both scenarios. The fault is simulated for two different grounding resistances, 10 and 0.1 Ω.

If a grounding grid is present there is only a small difference between both grounding resistors, being the magnitude of the AC component approximately the same in both. The DC component is larger for the smaller resistor, 45.5 A instead of 2 A, whereas the differences in the phase angle are smaller, because of the low magnitude of the current returning in the ground when compared with the magnitude of the current returning in the screen.

If there is not a grounding grid and the screen is connected to the ground, the magnitude of the AC current is larger for the lower grounding resistances, while the magnitude of the DC current is approximately the same for both grounding resistances. As a result, the relative DC current is larger for the larger grounding resistance, as expected. The phase difference between the voltage and the current is approximately 74° for the 10 Ω grounding resistance and approximately 27° for the 0.1 Ω grounding resistance.

It can also be concluded that to have a grounding grid is very similar to have a very low grounding resistance, if no grounding grid is present, as in this situation the current flowing in the ground is small when compared with the current flowing in the screens.[44]

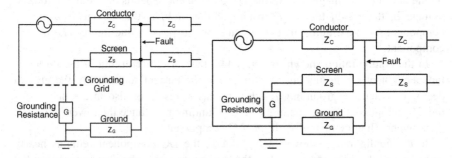

Fig. 4.44 Equivalent circuit for a single-phase cable during a fault

---

[44] This is assuming typical cases. It may be different if there are other conductors installed in the ground.

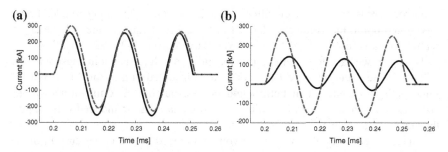

Fig. 4.45 Current in the conductor for different grounding resistances during a fault in a single-phase cable. *Solid line* 10 Ω grounding resistance, *dashed line* 0.1 Ω grounding resistance. **a** With grounding grid. **b** Without grounding grid

4.10.2 Three-Phase Cable

Having understood the specificities of the phenomenon for a single-phase cable, we take the next step and start the study of the phenomenon for three-phase cables.

A fault in a cable is typical to the ground.[45] The analyse of three-phase-to-ground fault is not very interesting for a transient point of view as the current and voltage in all three-phases decreases very rapidly to zero after the disconnection. Therefore, we analyse only a single-phase-to-ground fault (SPGF) as it allows the demonstration of the theory and its results can then be extended by the reader for two-phase-to-ground faults.

Consequently, we also do not analyse faults in pipe-type cables, as they will be similar to three-phase-to-ground faults.

Both-ends Bonding
Similar to what was done for a single-phase cable we can determinate the current return path during a fault. Figure 4.46 shows for a cable bonded in both-ends that the current returns through the ground, the screen of the faulted phase and the screens of the sound phases, because of the connection with the faulted phase at the end of the cable.

The large current in the conductor of the faulted phase induces a voltage in the sound phases, increasing the voltage in those phases, which will influence the TRV.

However, the voltage is not constant along the conductor. Considering a SPGF in the middle point of a three-phase cable, the magnitude of the voltage in the faulted phase decreases from the end(s) of the cable to the faulted point.

For simplicity, we say that one end of the cable is open and, thus, the current in the conductor is zero between the fault point and the open end. In this case, the voltage is not induced all along the sound phases, but only between energised end and the fault

---

[45] It is theoretical possible to have a phase-to-phase fault in a cable, as example for low voltages cables that are installed in a tunnel or a duct. However, it is very unlikely to have this type of fault for high voltage cables. Thus, we do not analyse them here.

point, as there is not any current after the fault. As there is no voltage induced after the faulted point, the voltage in the rest of the cable is equal to the voltage at the faulted location, unless when there is a very large current in the screens.[46]

However, we have to remember that the induced voltage is not in phase with the voltage of the sound phases. Thus, for a cable installed in a trefoil formation the same voltage is induced in both sound phases, but the peak magnitude of the voltages is different, as the total voltage in those phases is the induced voltage plus the steady-state voltage.[47]

It is also necessary to consider the voltage induced by the larger current in the screen. However, the current in the screens of the sound phases is approximately in phase opposition to the current in the faulted conductor and consequently it reduces the peak magnitude of the voltage in the conductor of the sound phases.

Summarising, the voltage in the sound phases continues to oscillate at 50 Hz, but the magnitude may increase because of the current in the faulted phase, which also oscillated at 50 Hz.

Having understood how the return current distributes itself by the phases and the ground during the fault, it is time to see what happens to the voltage waveform when the CB opens and clears the fault. Due to the complexity of the phenomenon, we need to use modal theory to explain the waveforms. For simplicity, we will initially consider only cases without grounding grid. This will not represent a lack of knowledge, as we already know that the grounding grid cases are very alike to cases where the grounding resistance is very low and both are considered equivalent in the analysis done next.

Screenless cable

We start by considering a cable consisting only in a conductor and insulation, i.e., no screen. This situation, which we know as unlikely to happen for HV, may occur for lower voltages and it is a good way of introducing the waveforms.

The fault to ground is in the conductor of phase A, whereas the other two phases remain sound. For simplicity, we assume that the phase A is the last one to be switched-off.

When the CB opens in the faulted phase, a voltage transient wave is generated as explained before for the single-phase example. By electromagnetic coupling, voltage waves are also generated in the other two conductors, which superimpose to the waves already present in those conductors.

At this point, we introduce the modal domain. We have three conductors and the ground, meaning that we have two conductor modes[48] and a ground mode. By

---

[46] Some simplifications are being made here, most notability we are not considering the voltage induced by the current in the screens of the sound phases, which also increases during the fault. We are going to see later that for low grounding resistances and/or existence of grounding grids the voltage changes after the fault point.

[47] Download and run the PSCAD files available online in order to observe this difference in detail.

[48] One of modes is an interconductor mode, which is not influenced by the transient.

applying the methods explained in Sect. 3.4, we obtain the voltage transformation matrix of the system (4.53), where the first column is the ground mode, the second the conductor mode and the third is the interconductor mode.

$$[T_V] = \begin{bmatrix} 1/\sqrt{3} & 2/\sqrt{6} & 0 \\ 1/\sqrt{3} & -1/\sqrt{6} & -1/\sqrt{2} \\ 1/\sqrt{3} & -1/\sqrt{6} & 1/\sqrt{2} \end{bmatrix} \tag{4.53}$$

The interconductor mode is zero to phase A and it does not influence the phenomenon. The ground mode is equal in all three phases in magnitude and polarity. The conductor mode shows that the transient wave generated in phase A will have half of the magnitude and opposite polarity in the other two phases.

We also know that the velocity of the conductor and ground modes are different, meaning that there will be two waves flowing at different speeds in the conductors.

What we do not know is the magnitude of each mode. We are analysing a SPGF, thus, we can expect to have a ground mode with a large magnitude. The demonstration can be made using the inverse of the transformation matrix (4.54) and (4.55).

$$[T_V]^{-1} = \begin{bmatrix} 1/\sqrt{3} & 1/\sqrt{3} & 1/\sqrt{3} \\ 2/\sqrt{6} & -1/\sqrt{6} & -1/\sqrt{6} \\ 0 & -1/\sqrt{2} & 1/\sqrt{2} \end{bmatrix} \tag{4.54}$$

$$[T_V]^{-1} V_P = V_M \tag{4.55}$$

We have seen that the voltage in the sound phases changes during the fault. Thus, at the disconnection instant, the magnitude of the voltage in those phases is no longer approximately 1 pu. The question now is if these voltages are larger or smaller than 1 pu. The voltage difference between the current and the voltage of the faulted phase is in theory between 0 a 90° depending on the value of ground resistance. However, the typically values can be seen as being between 30° and 60°.[49] The current induces a voltage in the sound phases with a polarity opposite to derivative of the current. Thus, the voltage induced in the sound phases has a phase difference between 120° and 150° to the voltage of the faulted phase. The phase difference between the induced voltage and the voltage of the sound phases will then be between 0° and 30° for one phase and 90° and 120° for the other phase. For those values and usual coupling factors, the voltage in the sound phases will typically increase.

The voltage in the sound phases increases between the sending end, which we consider connected to an ideal voltage source, and the faulted point, after that point there is no current in the conductor and thus, no voltage is induced in the sound phases. We have also seen for the single-phase example that at the disconnection

---

[49] These values have already some tolerance and for real cases the interval is even more narrow.

moment the charges in the cable need to balance, resulting in the generation of an impulse voltage.

Using this information together with (4.54) and (4.55), we can understand why the magnitude of the ground mode is usually larger than of the conductor mode. The conductor and ground modes are calculated as shown in (4.56). The magnitude of ΔV_{PhA} is approximately 1 pu, whereas the magnitude of ΔV_{PhB} and ΔV_{PhC} depends on the layout of the system, but they are usually less than 1 pu. For simplicity, we also consider the latter two as being equal. It is easy to see that in these conditions the magnitude of the ground mode will be larger than of the conductor mode.

$$\begin{cases} M_C = \Delta V_{PhA} \frac{2}{\sqrt{6}} - \Delta V_{PhB} \frac{1}{\sqrt{6}} - \Delta V_{PhC} \frac{1}{\sqrt{6}} \\ M_G = \Delta V_{PhA} \frac{1}{\sqrt{3}} + \Delta V_{PhB} \frac{1}{\sqrt{3}} + \Delta V_{PhC} \frac{1}{\sqrt{3}} \end{cases} \qquad (4.56)$$

Having laid down the theoretical foundations of the phenomenon, we can finally see the waveforms. Figure 4.47 shows the voltage in all three phases for two points of the cable. The waveforms on the left are for voltage at the sending end of the cable, while the ones on the right are for a point at an equal distance from the sending end and the fault.

The presence of the modes can be seen on the second figure. At approximately 0.05 ms (point A), there is an increase of the voltage in phase A, while the voltage in the other two phases decreases. It is also observed that variation in Phase A is approximately the double of the variation in the other two phases. Thus, it is concluded that these variations are associated to the conductor mode.

At approximately 0.1 ms (point B), the slope of phase A increases, while the voltage in the other two phases starts to increase. Thus, it is concluded that the ground mode voltage as arrived to the point at this instant. As the two mode voltages have different speeds, the waveforms have the distorted shape shown in the figures. The distortion is larger in phases B and C because the two modes have different polarities for these two phases, the ground mode increases the voltage, whereas the conductor mode decreases it.

A consequence of this transient is that the voltage at the terminals of the CB may reach rather high values for phases B and C than for a normal disconnection. In the example shown in Fig. 4.47, the voltage at the terminals of the CB will be larger than 2 pu for phases B and C, as the voltage in these two phases is larger than 1 pu half-cycle after the disconnection. Moreover, the slope of the TRV is larger than for a normal de-energisation increasing the risks of a reignition/restrike.

Cable with screen
The analysis just done was for a cable without a screen. The next step is to see what happens in the same situation for a cable with screen bonded at both-ends.

Like before, it is necessary to know the voltage distribution along the cable before the switch-off, in order to know the magnitude of the transient waves. We have seen the existence of a grounding grid, or the value of the grounding

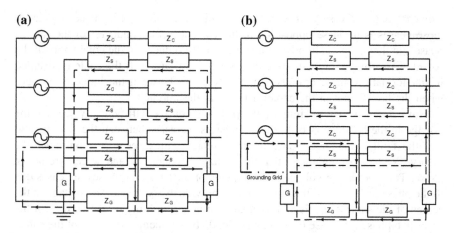

Fig. 4.46 Equivalent circuit for a three-phase cable bonded in both-ends during a single-phase-to-ground fault. **a** Without and with grounding grid

resistance if one is not present, are very important and influences the currents during the short-circuit, meaning that the induced voltages are also influenced.

As it was previously explained, the return current divides itself by the screens and grounds (Fig. 4.46) depending on the value of grounding resistance (remember that a very low grounding resistance is equivalent to a grounding grid). If this resistance is large, the current returning in the screens is small when compared with the current in the conductor of the faulted phase, as much of the current returns in the ground. Thus, the majority voltage induced in the screen of the sound phases during the fault is generated by mutual coupling with the conductor of the faulted phase, which has a large current. Assuming that the cables are laid close, as it is many times the case, the voltage generated in all three screens is approximately the same, being slightly larger in the screen of the faulted phase, both because of the proximity to the conductor and the larger current in that specific screen. The same reasoning is applied to the conductors of the sound phases.

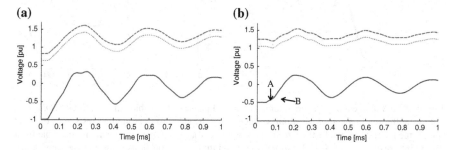

Fig. 4.47 Voltage in the three phases after the CB switch-off. *Solid line* Phase A, *dashed line* Phase B, *dotted line* Phase C. **a** Voltage at the sending end of the cable. **b** Voltage at the middle point between the sending end and the fault

However, it is necessary to remember in this case the current in the screens of those phases is in phase opposition to the current in the conductor of the faulted phase and they help to slightly reduce the voltage induced in the conductors of the sound phases by the faulted phase during the short-circuit. If the cable is installed in flat formation, there are also some differences between the induced voltages when the fault is not in the cable installed in the middle.

However, if we try to simulate the phenomenon we will typically see that the voltage in the conductors of the sound phases increases between the sending end and the fault, while decreasing in the screens, which seems to contradict the theory. We have to remember that the voltage in conductor and screen are not in phase. Thus, while the voltage in the conductor of the sound phases increases due to the phase difference to the faulted phase previously explained, the voltage in the screen of the same phase is approximately in phase with the screen voltage of the other two phases, because of the connection between them at the ends of the cable. As a result, the voltage in the screens is in phase with the voltage in the conductor of the faulted phase and the induced voltage will be in phase opposition.

For large grounding resistance, the magnitude of the voltage is large at the sending and as a result the voltage decreases between the sending end and the fault. If the voltage reaches zero before the fault, the magnitude starts to increase, but with opposite polarity. For a low grounding resistance, the voltage at the sending end is close to zero and the voltage increases between the sending end and the fault with opposite polarity. An example of this is shown in Figs. 4.48 and 4.49.

Low grounding resistors or grounding grids

The previous description is valid for large grounding resistance, but not for low grounding resistances or systems where a grounding grid exists. In this case, the returning current in the screens increases, being larger and with an order of magnitude closer to the current in the conductor of the faulted phase, which also increases, as the total impedance of the system decreases. As a result, the voltage

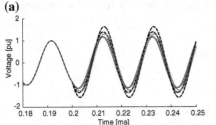 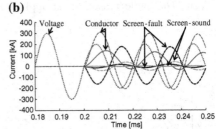

Fig. 4.48 a Voltage along the conductor of a sound phase of the cable during a SPGF. *Solid line* voltage at the sending end, *dashed line* voltage along the cable for a grounding resistance of 10 Ω (larger the magnitude, the furthest from the sending end), *dotted line* voltage along the cable for a grounding resistance of 0.1 Ω (larger the magnitude, the furthest from the sending end). **b** Currents in the conductor of the faulted phase, current in the screens and voltage (not at scale) in the faulted phase during a SPGF. *Solid lines* grounding resistance of 10 Ω, *dashed lines* grounding resistance of 0.1 Ω

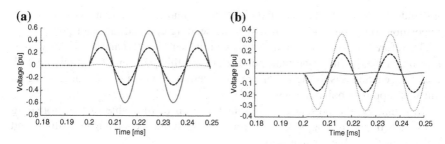

Fig. 4.49 Voltage along the screens of a sound phase of the cable during a SPGF. **a** Grounding resistance of 10 Ω. **b** Grounding resistance of 0.1 Ω. *Solid line* end of the first section, *dashed line* end of the second sections, *dotted line* end of the third section

induced in the conductor and screen of the sound phases depends also of the current in the screens, which can no longer be overlooked.

In this case it is difficult to predict the magnitude and phase of the waveforms in the sound phases. For simplicity, we continue to assume a fault at the middle of the cable. The current in the conductor of the faulted phase has a high magnitude between the sending end and the fault, being zero after that point.

The magnitude and phase of the current in the screen of the faulted phase change at the fault point. The current returning in direction to the sending end is approximately in phase opposition to the current in the conductor, whereas the screen's current flowing to the receiving end is approximately in phase with the current in the conductor. The latter will then return in the screens of the sound phases.

We now start to analyse the voltages from the sending end to the receiving end, assuming the same referential direction for the currents. We overlook the current in the conductors of the sound phases for being too low, when compared with the other currents. Thus, the only currents considered are the ones in the screens of the three phases, plus the current in the conductor of the faulted phase. The current in the three screens are approximately in phase, between the sending end and the fault, whereas the current in the conductor of the faulted phase is approximately in phase opposition to these three currents. The magnitude of the current in the conductor is larger than the summation of all three screen currents.[50] Thus, the voltage induced in the screens and conductors of the sound phases typically increases as explained for the screenless cable.

After the fault point, the current in the conductor of the faulted phase is zero and there is only current in the screens. There is also current flowing through the ground to the receiving end and then flowing back to the sending end through the screens. In this case the summation of the current in the screens of the sound phases is larger than the current in the screen of the faulted phase. Consequently,

---

[50] Remember that part of the current returns in the ground.

the induced voltage between the fault point and the receiving end is phase opposition to the one induced between the sending end and the fault point.

Therefore, the voltage increases between the sending end and the faulted point reaching a maximum at the fault point. Between this point and the receiving end, the voltage decreases. If the voltage reaches zero it will then start to increase again but with opposite polarity to before.[51]

In summary, the presence of a grounding grid or the value of the grounding resistance if one is not present, have a strong influence in the magnitude and polarity of the voltage waveform.

However, there are some aspects that can be expected in advance:

- The voltage in the conductor of the sound phases increases between the sending end and the fault point. Depending on the ground resistance, the voltage remains approximately constant after this point (for large grounding resistance) or has a slight decrease (for very low grounding resistance or grounding grids);
- The voltage in the screen of the sound phases seems at first more unpredictable, but in reality it is similar. The difference is that the relative variations are much larger and as a result the voltage in both-ends of the screens may even be in phase opposition. However, it is valid the same physical explanation;
- We have seen for the single-phase that there is a phase difference between the voltage and current of the faulted phase, which is between 30° and 60°. Typically, the voltage induced by this current will increase the voltage in the sound phases during a fault;

Figure 4.48 shows voltages along the conductor of a sound phase and the currents in the conductor of the faulted phases and all the screens during a SPGF for two different grounding resistances. It is possible to observe that the voltage increases more when the grounding resistance is larger and that the phase difference between the voltage and current is larger when the grounding resistance is larger, because of the larger relative current flowing in the ground. As expected, the faulted current is larger for lower grounding resistances, as it is the current in the screens, which can virtually be overlooked for large grounding resistances.

For a case with a grounding grid it would be similar to a low grounding resistance, but the currents would be slightly larger because of the lack of the 0.1 Ω impedance between the screens and the common point.

Figure 4.49 shows the voltage along the screens of a sound phase during a SPGF for two different grounding resistances. It is possible to see that for the large grounding resistance the voltage in the screen decreases when going form the sending end to the fault, whereas the opposite happens for the low grounding resistance, as previously explained.

Using this information together with the inverse voltage transformation matrix of a trefoil cable (4.57), we can try to predict how the transient voltage is after the disconnection of the CB.

---

[51] It is virtually impossible for the conductors, but it may happen for the screens.

$$[T_V]^{-1} = \begin{bmatrix} 0 & 0 & 0 & 2/\sqrt{6} & 2/\sqrt{6} & 2/\sqrt{6} \\ 0 & 0 & 0 & 0 & 1 & -1 \\ 0 & 0 & 0 & 2/\sqrt{3} & -1/\sqrt{3} & -1/\sqrt{3} \\ 1/\sqrt{3} & 1/\sqrt{3} & 1/\sqrt{3} & -1/\sqrt{3} & -1/\sqrt{3} & -1/\sqrt{3} \\ 0 & 1/\sqrt{2} & -1/\sqrt{2} & 0 & -1/\sqrt{2} & 1/\sqrt{2} \\ 2/\sqrt{6} & -1/\sqrt{6} & -1/\sqrt{6} & -2/\sqrt{6} & 1/\sqrt{6} & 1/\sqrt{6} \end{bmatrix} \quad (4.57)$$

We start by considering an ideal voltage source and a large grounding resistance, which means that the voltage is approximately equal in all three screens and that it increases between the sending end and the fault point for the conductor of the sound phases, remaining constant after that point. The voltage profile along conductor of the faulted phase is like for the other cases, 1 pu at the sending end decreasing along the cable and reaching zero at the fault.

We can immediately say that the intersheath modes are approximately zero and are not influence by the switch-off. We also know that three screens are connected to each other, meaning that all three will be connected to the fault. Consequently, the energy in the screens will damp very rapidly through the fault and the ground mode will disappear after some milliseconds.

The variation of the voltage in the conductors is a result of the balance of the charges according to the voltages' profiles previously described. For large grounding resistance, the voltage induced in the conductors of the sound phases is similar to the voltage induced in the screens.[52] Thus, the variation of the voltage magnitude between the sending end and the fault is approximately the same for the conductors and screens of the sound phases. As a result, one of the conductors modes remains constant after the switch-off as the summation of all components is close to zero, whereas the other two conductor modes depend only on the voltage profile in the conductor and screen of the faulted phase.

Consequently, the variations on the voltage magnitudes after the disconnection depend mainly on the ground mode.[53] As this mode is damped very fast, the transient associated to the short circuit also disappears very fast and after that the voltage profile resembles a normal cable de-energisation.

The question now is if the voltages associated to the ground mode will increase or decrease the voltage in the sound phases. We have seen that the voltage in the screens is an induced voltage and how the induced voltage increases the magnitude of the voltage in the sound phases. Consequently, we can conclude that the voltage associated to the ground mode will increase the voltage in the sound phases after opening the CB.

---

[52] Remember that the magnitude of the current in the screens is rather low for large grounding resistances.

[53] The voltage variation is larger for the conductor of the faulted phase (1 pu) than for the screen (≤ 1 pu). As a result, there is a small initial variation in two of the coaxial modes. However, the larger the grounding resistance the larger the voltage variation in the screen and in the limit case the voltages are almost equal (assuming that the cables are not too far apart from each other).

It is now time to go to the other extreme case and see what happens when the ground resistance is very low or a grounding grid is installed, i.e., the more realistic case.

We know that the voltages in the screens of the sound phases are similar. Consequently, we can immediately say that one of the intersheath modes is zero. The remaining modes are a little more complicated of analysing and several situations may happen depending on where in the cable the fault occurs.

However, we can make some simplifications without a substantial loss of accuracy. For very low grounding resistances, the voltage at both-ends of the screens is close to the ground potential. As the voltage in both-ends on the screens is similar and low, their transient after the switch-off has a very low magnitude and can be discarded, without loss of accuracy. Thus, we can say for this case that the ground mode and the two intersheath modes do not have a substantial influence the transient.

We have seen before that the lower the grounding resistance, the lower the phase difference between the voltage and current in the faulted phase. Thus, the instantaneous voltage at the sending end of the faulted phase at the disconnection moment is lower than for large grounding resistances.

Another consequence of the phase difference is that the superposition of the induced voltage in the sound phases does not result in the same voltage waveform for both sound phases, as explained for the screenless cable. Thus, one of the sound phases has typically a larger voltage variation between the sending end and the fault point than the other.

Additionally, for large grounding resistances, the current in the sound phases has a large variation when the faulted phase is disconnected. This variation is usually large enough to make this currents cross zero, allowing the immediate disconnection of the sound phases. The same is not true for low grounding resistances, where there is only a small variation in the magnitude of the current at the sound phases. Thus, it is common to have the sound phases still connected for some milliseconds after the disconnection of the faulted phase, assuming that the faulted phase is disconnected first.

Therefore, it is not easy to predict how the voltage will behave. However, if we put together all that we know we can conclude that in principle the voltage will not increase as much as for large grounding resistances. The voltage variation at the disconnection instant is smaller for small grounding resistances both for the faulted phase, because of the lower phase difference, and sound phases because of the larger current in the screen of the faulted phase that induces a voltage with opposite polarity to the one induced by current in the conductor. As a matter of fact, in some cases the voltage even has a small decrease after the switch-off.

Figure 4.50 shows the phase and modal voltage during the disconnection for a very large and a very low grounding resistance. It is possible to see how the voltage increases in the sound phases after the disconnection for large grounding, but not so much for the low grounding resistance, where there is even a small voltage decrease in one of the phases. One of the sound phases is disconnected before the faulted phase and the voltage in that phase increases because of the

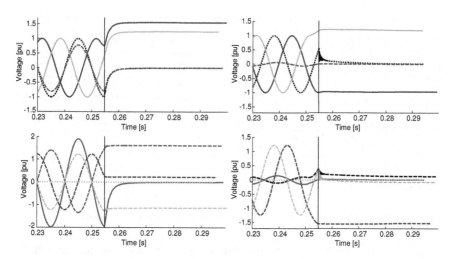

Fig. 4.50 Example of the phase (*up*) and modal (*down*) voltages at the sending end of the cable before and after the disconnection for a very large (*left*) and a very low (*right*) grounding resistance. *Up solid lines* voltage in the sound phases, *dotted line* voltage in the faulted phase, *dashed line* voltage in the screen. *Down solid line* ground mode, *dotted line* intersheath modes, *dashed lines* coaxial modes

coupling with the faulted phase, we can see that the voltage stops to increase when the faulted phase is disconnected.

It is also possible to observe the modal voltages and see how the ground mode is the only mode varying for a large grounding resistance and how for a low grounding resistance there is only a small variation in two of the coaxial modes.

Cross-bonded cables
The previous explanations were for a cable bonded in both-ends. Would there be changes if the same cable was instead cross-bonded?

A cross-bonded cable has a lower series resistance than an equivalent cable bonded in both-ends, both during steady-state operation and during a fault (for a fault in the same point of the cable). Figure 4.51 shows a possible SPGF in a cross-bonded cable with two major-sections and some of the return paths for the current. In this case, the fault is considered as being in the first major-section.[54] For a system with a grounding grid, the figure would be similar but the current returning in the screen would flow to the source without passing by the resistor, whereas the current flowing in the ground would pass by the resistor, see Fig. 4.46 for the cable bonded at both-ends example.

Part of the current returns in the screen of the faulted phase like with a cable bonded in both-ends, the difference is that the screen is connected to the sound phases, which will influence the waveforms in those phases because of mutual

---

[54] For simplicity we do not consider the link boxes and their impedances.

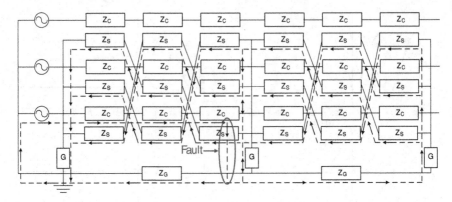

Fig. 4.51 Fault in a cross-bonded cable and some of the current's return paths

coupling. The current that is returning directly to the sending end through the ground is also alike the cable bonded in both-ends.

The main difference is in the other direction. In a cable bonded at both-ends, part of the current would flow to the receiving end in the ground and screen of the faulted phase until the grounding point at the receiving end and then return to the sending end in the screens of the sound phases. In this example, there is a grounding point closer to the fault and part of the return current flows in the ground and screen of the faulted phase up to that point and returns to the sending end in the screens of the sound phases, no longer needing to flow all way until the end of the cable.[55] Consequently, the equivalent impedance is lower and the magnitude of the short-circuit current is larger.

Even if the fault was in the second major-section, the current returning to the sending end through the ground and screen of the faulted phase would divide itself between the ground and the screens at the grounding point in between, reducing the resistance of the system when compared with a cable bonded in both-ends.[56]

By other hand, if the cable has only one major-section, the differences for an equivalent cable bonded in both-ends are minor and resultant of the changes mutual coupling between the screen and the conductor resultant of the transposition of the screens.

Having understood how the current behaves during the fault is time to see what happens to the voltage. We saw for the cable bonded in both-ends how important is the coupling between phases, because of the high currents associated to the

---

[55] Part of the current continues flowing in the ground until the end of the cable. The division of the current depends on the impedance values as given by classic circuits theory. It would be the same for cables with several major sections.

[56] As most of the transient phenomenon, a proper simulation of a fault requires an accurate model. A common error made when simulating this phenomenon is the grounding. The connection of the screens should reflect the real system, typically all three phases are connected together and there is a common resistance for all three. Depending on the software, one should also be careful with the grounding when dividing the projects by different working modules.

phenomenon. The question now is how the transposition of the screens affects the results.

We have seen that for large grounding resistances, the magnitude of the current in the screens is rather low. Thus, the cross-bonding of the screens barely affects the results and the voltage waveforms are almost equal to the ones that would be obtained for an equivalent cable bonded in both-ends and a fault in the same point.

The same will not be true for low grounding resistances, where there is more current flowing in the screens. We consider a fault in the location shown in Fig. 4.51 for a cable with two major-sections and a fault in the same location for a cable with only one major-section. We know that the short-circuit current is larger for the cable with two major-sections because of the lower resistance faced by the return current.

However, the DC current, if present, is similar for both configurations, as it depends on the ground, meaning that the phase difference between the voltage and the current is also similar for both. Moreover, the DC component current will be small when compared with the AC component and not very relevant for the analysis.

The current in the screens is also larger for the cable with two major-sections than the one with one major-section. Except for the path between the fault and the sending end, in that case the magnitudes are similar for both configurations as the impedance is also similar for both.

Knowing this, we can extrapolate on how the voltages increase along the cable during a SPGF. By one hand the cable with two major-sections has a larger current in the conductor, which will induce a voltage in the other conductors, by other hand it has also a larger current in the screens, which induces a voltage with opposite polarity to the one induced by the conductor limiting the voltage increase.

However, the current in the conductor is larger than the summation of the current in the screens. Consequently, the voltage in the sound phases increases typically more for the cable with two major-sections than for the cable with one major-section. This is generally true, but it also depends on where the fault occurs. Considering a fault as shown in Fig. 4.51 for both cases, the voltage variation in one of the sound phases of the cable with one major-section would be large for one phase, but not for the other. The reason is that for fault in that point, approximately at middle, the return current flowing from the fault to the sending end does not flow in the screen of the upper cable when having only one major-section. Therefore, the voltage induced in this phase by the screen's current is lower and the majority of the induced voltage is from the current in the faulted conductor. As a result, the voltage in that phase is larger for the cable with one major-section than for an equivalent cable with two-major-sections. Yet, it is obvious that this is very dependent on where in the cable the fault occurs and each case is a case.

Wrapping all together, we see that the behaviour would be similar to the one seen for a cable bonded in both-ends, but with some differences because of the differences in the magnitudes of the voltages.

4.10.3 Short-Circuit in a Cable Connected to a Shunt Reactor

We have previously seen how the de-energisation of a cable together with a shunt reactor changes the waveforms, which are no longer a decaying DC, but a decaying AC oscillating at resonance frequency.

The same happens for the sound phases when the CB opens because of a fault. The question now is to know if the magnitude of the current and voltage are affected by the shunt reactor. The answer depends on where the shunt reactor is installed, its impedance and the magnitudes of both voltage and current at the disconnection instant.

We start by analysing how the waveforms are affected during the fault prior to the disconnection of the CB. The voltages and currents during the fault are similar, being slightly lower in the system with the shunt reactor. The difference is noticed in the conductor of the sound phases, but it is too small to influence the magnitude of the short-circuit current. Thus, the voltage in the sound phases is approximately the same with or without shunt reactor. This is valid independently of where along the cable the shunt reactor is installed.

In order to explain the behaviour of the waveforms after the disconnection, we are going to analyse a simplified version of the circuit first. Figure 4.52 shows a single-phase circuit that can be seen as a simplified version of one of the sound phases immediately after the disconnection. Where, C is the cable capacitance, R is the cable resistance and L_S is the shunt reactor inductance.

The mathematical solution of the circuit shows that theoretical maximum magnitude of V_1 is given by (4.58).

$$V_1 = \sqrt{V_{Avg}^2 + V_{I\_Sh}^2} \tag{4.58}$$

$$V_{Avg} = \frac{V_1(0) + V_2(0)}{2} \tag{4.59}$$

$$V_{I\_Sh} = I_{Sh}(0)\sqrt{\frac{L}{C}} \tag{4.60}$$

Fig. 4.52 Single-phase equivalent circuit immediately after the disconnection

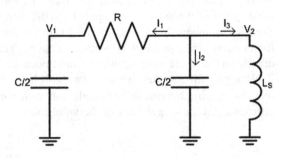

Equation (4.58) shows that magnitude of the voltage in the circuit after the disconnection depends on the inductance of the shunt reactor and the disconnection instant. The two extreme cases would be to disconnect the circuit when the voltage is at a peak value, and thus the current in the shunt reactor is zero, or do the opposite and disconnect the circuit when the voltage is zero and the current in the shunt reactor is at a peak value. In reality, we are limited by the fact that a typical CB opens only for 0 A.

The equation continues to be valid for the shunt reactor installed in the other side of the circuit at the voltage point V_1. This is an important result as it allows us to conclude that the location of the shunt reactor influences the magnitude of the waveform after the disconnection.

This can be demonstrated by considering that the magnitude of V_2 is larger than the magnitude of V_1.[57] This means that except for the extreme case when the current in the shunt reactor is 0 A at the disconnection instant, the current $I_{SH}(0)$ of (4.60) is always larger if the shunt reactor is installed in the voltage point V_2. Consequently, we can conclude that voltage after the disconnection is larger if the shunt reactor is installed at the voltage point V_2.

We can export this result for the case where the shunt reactor is installed in a cable and use it there. We have seen that sometimes that voltage in the sound phases increases along the conductor. Thus, we can normally expect to have larger overvoltages if the shunt reactor is installed in a point where the voltage is larger at the disconnection instant. In the examples that we used before, would be if the shunt reactor was installed the furthest away from the sending end, in a more realistic case it would if the shunt reactor was installed at the end that is disconnected first.[58]

We have seen that the location of the shunt reactor influences the waveform, but we have not yet compared to an equivalent system without shunt reactor. Equation (4.60) tells us that the larger the inductance of the shunt reactor, the lower the voltage after the disconnection, as the value of $I_{Sh}(0)$ is inversely proportional to L. As an infinite of L is equivalent to not having a shunt reactor, the conclusion would be that the shunt reactor would increase the overvoltage.

However, a realistic system is more complex and the reality is that we cannot say what will happen without doing a simulation. As an example, we have to remember that the shunt reactor also has resistance, which increases the damping; the shunt reactor also changes the phase angle between the voltage and the current, which we have seen to be very important. Thus, in some situation the voltage will be larger if a shunt reactor is presence in other situations it will be the opposite.

---

[57] This is the main difference for a normal de-energisation where V_1 and V_2 have similar values.

[58] Remember that during the fault the voltages increases along the conductor of the sound phases from the energised end to the open end.

Mutual coupling

The previous analysis was done without considering mutual coupling between the phases of the shunt reactor. If we consider mutual coupling there is an interesting phenomenon that may happen for SPGF and lead to a larger overvoltage.

To demonstrate the phenomenon, we assume that a fault occurs very close to the shunt reactor in phase A. In this situation, the current into the faulted phase of the shunt reactor is a DC current with derivate equal to zero. As a result, Phase A does not induce voltage in the other two phases, creating an asymmetry between the shunt reactor phases, which is responsible for a voltage increase in the sound phases.

This means that the higher overvoltage is not a direct consequence of the fault, but the result of not having phase A induces voltage in the other two phases. In other words, the higher overvoltage obtained for the SPGF is equal to the overvoltage obtained if there were only mutual inductance between phases B and C. A corollary of these results is that this phenomenon can only be present for a SPGF. If there is a fault to ground for more than one phase, there will be a DC current in more than one phase and no mutually induced voltage.

The reasoning can be also applied if the fault is located some kilometres away from the shunt reactor, with the difference that some current is flowing in the faulted phase. Whereas, the current is very low, barely inducing any voltage in the other two phases.

However, we should remember that this voltage increase depends on several factors like the mutual inductance value and the time between the disconnection of the phases (see Sect. 4.6).

Figure 4.53 exemplifies the phenomenon by showing three different scenarios. In all three scenarios, a shunt reactor is connected at the sending end of the cable and it is de-energised together with the cable. In Fig. 4.53a it is shown a normal de-energisation, i.e., there is no fault, for a shunt reactor with a mutual coupling of $-0.05H$ between the phases. In Fig. 4.53b there is SPGF, but no mutual coupling in the shunt reactor. Finally in Fig. 4.53c, there is a SPGF and mutual coupling in the shunt reactor. It is observed that the TOV is larger and longer when there is mutual coupling between the phases of the shunt reactor.

4.10.4 Influence in the Short-Circuit of Other Network Equipment

We have seen the voltage and current waveforms during a SPGF for cables directly connected to an ideal voltage source. However, a fault in a cable connected to a transformer is more likely to happen.

We analyse a short-circuit of a cable connected to a transformer, where the CB is installed between the cable and the transformer, as it is usual. For our first analysis, we consider the transformer as having a star–star connection with both

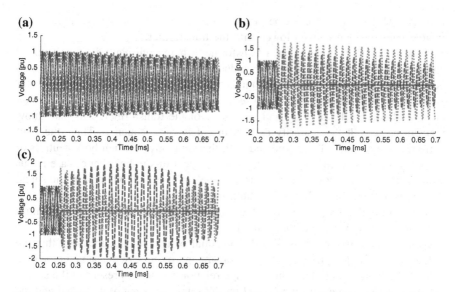

Fig. 4.53 Voltage at the sending end of the cable after the disconnection for different configurations. **a** Shunt reactor with mutual coupling, but no fault in the cable. **b** Shunt reactor without mutual coupling and fault in the cable. **c** Shunt reactor with mutual coupling and fault in the cable

neutrals connected to earth, with an ideal voltage source in the source side and without saturation. As before, we do the analysis for a SPGF and the results for other types of faults can be extrapolated from this analysis.

This circuit can be seen as an equivalent to have a big inductor between the cable and the source. Thus, the current during the short-circuit is smaller than when the cable is connected to an ideal voltage source. Consequently, the current in the screens during the fault is also smaller. As the current in the faulted is smaller, the voltage induced in the sound phases, both conductors and screens, is also lower.

Another difference is in the voltage at the sending end of cable for the faulted phase. In the previous example, the voltage at the sending end was the voltage of the source. In this case as there is an inductance between the source and the cable, the voltage at the sending end of the faulted phase is much reduced. As a result, the voltage variation between the sending end of the cable, where the CB is installed, and the faulted point is lower, resulting also in a lower variation of the voltage after the opening of the CB.

Putting all this information together, we can conclude that the voltage increase in the cable after the disconnection is lower when the cable is connected to a star–star transformer with both neutrals grounded. Remember that for the large grounding resistances the voltage increase was associated to the ground mode, which is depended on the screen voltages that are now lower. For low grounding resistances or grounding grids, we saw that one of the reasons for a lower voltage increase, which may not even be present, was the lower voltage variation in the faulted phase, which is even lower with the transformer.

This explanation is valid for a transformer where both neutrals are grounded and the transformer can be seen as an inductor even during the fault. The next step is to see what happens if one of the neutral is not grounded in the star–star connection. It is known that in these conditions zero-sequence current cannot circulate between the primary and secondary of the transformer. It is also know that during a fault the majority of the current is zero-sequence. Thus, the fault current is very small when compared with the fault current when both neutrals of the transformer are grounded. So, we would expect this to be a better case for us with lower currents, however, there is a "problem" with the voltage of the sound phases, which increases during the fault, increasing also the TRV voltage.

We assume that the load side of the transformer is the one that has the neutral grounded. When we have the fault, the voltage in the faulted phase goes to approximately zero, which is also the potential of the ground, meaning that the voltage of the faulted phase and the neutral are approximately the same in the load side of the transformer. The same voltage relation has to be present in the source side of the transformer. As the neutral on the source side is not grounded and the transformer is considered to be connected to an ideal voltage source, the voltage of the neutral increases and becomes equal to the source voltage for the phase that is short-circuited in the load side. Consequently, the phase-to-neutral voltage in the sound phases increases and becomes equal to phase-to-phase voltage of the balance system, approximately 1.73 pu.

In the previous example where both neutral were grounded, it was not possible for the voltage in the neutral to become equal to the phase voltage. Thus, the magnitude of the voltage in the sound phases in the sound is almost equal to the magnitude in normal operation conditions.[59] Another consequence of this is that the voltage in the faulted phase in the load side of the transformer is not zero, however, it is smaller than prior to the fault.

If the neutral is grounded in the source side of the transformer, the voltage for the faulted phase reduces to virtually zero in the load side. However, the system continues balanced in the source side, thus, the voltage in the neutral of the load side is in phase opposition to the phase in the source side that is equivalent to the faulted phase. Thus, the voltage of the sound phases in the load side is again approximately equal to the phase-to-phase voltage in normal operation conditions.

At last we analyse the most typical power transformer connection, the star-delta connection with grounded neutral, starting with the currents. The zero-sequence impedance depends from which side of the transformer it is seen. If seen from the star side the zero-sequence impedance is the short-circuit impedance as the zero-sequence current can circulate to the ground. If seen from the delta side, the zero-sequence impedance is virtually infinite as there is no circulation path for the zero-

---

[59] Remember that the current in the screens and conductors is low and there is almost no voltage increase because of mutual coupling.

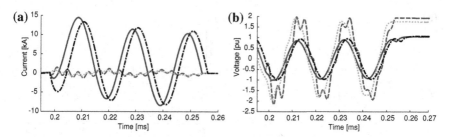

Fig. 4.54 Waveforms during and after clearing a SPGF. **a** Current waveforms in the faulted phase at the cable sending end. **b** Voltage waveforms in a sound phase on the load side of the CB; *Solid line* Yn/Yn, *dashed line* Yn/Y, *dotted line* Yn/Δ, *dashed-dotted line* Δ/Yn

sequence currents. Therefore, it is not the same to have a star-delta connection or a delta-star connection, being the short-circuit current larger if the cable is installed in the star side of the transformer. In this last case, the currents in the load side of the transformer are similar to those obtained for a star–star connection where both neutrals are grounded.

The explanations previously given for the voltage continue to be valid. If the cable is connected to the delta side, the voltage in the faulted phase is reduced to zero and the magnitude of the voltage in the sound phases increases to approximately to the typical phase-to-phase value (1.73 pu). If the cable is connected to the star side, the voltages are similar to those of the star–star connection with both neutrals grounded.

A final word about the influence of saturation, we considered the transformer connected to an ideal voltage source and thus the saturation had little influence in the waveforms. In more weak networks, this would no longer be true.

Figure 4.54 shows the current and voltage waveforms during and after a SPGF for different types of transformer connections. It is possible to see that the current in the faulted phase is larger when there is a neutral star connection on the load side of the transformer and that the voltage is sound phases is larger when there is a delta connection or an ungrounded star connection on the load side of the transformer.

The harmonics seen in some of the waves are dependent of the current and voltage values at the fault instant and the interaction between the inductive and capacitive elements of the system.

Another factor that we have to consider when studying this phenomenon is the capacitance to the ground of the transformer's windings and bushing.

Figure 4.55a shows a simplified version of a circuit during a fault, where the inductor represents the transformer and the capacitor represents the capacitance to ground in the transformer. During the fault, the voltage in the faulted phase is zero at both sides of the CB is zero while it is close. When the CB opens the voltage in the load side of CB remains zero, but the voltage at the capacitor tends to equalise the source voltage, which is not zero at the disconnection instant, initialising a

(a)   **(b)**

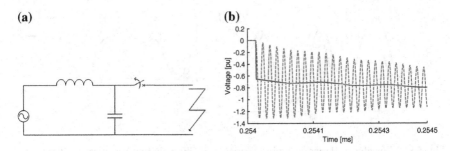

Fig. 4.55 Simplified circuit for the analysis of a TRV and correspondent simulation voltage. *Solid line* without ground capacitor, *dashed line* with ground capacitor

transient whose waveforms were explained in Chap. 2. As a result, the TRV is sharper and the probability of a reignition is larger.

This reasoning is valid also for normal disconnection or for when there is not a transformer. However, the capacitance to ground in the source side of the CB is substantially lower if there is not a transformer in that location.

Figure 4.55b shows the TRV voltage at the faulted phase with and without a capacitor to ground on the source side of the CB for a star–star connection with both neutrals grounded. The transient in the source side of the CB can be seen for the model with capacitor, but not for the model without capacitor.

However, there is an extra aspect that we have to talk about that is not seen in the simulations shown in Fig. 4.55b. In the simulation the CB is model as an ideal CB, consisting in a very small resistance when close and very large resistance when open. In a more realistic model there would be also a capacitor between the terminals of the CB and a capacitor to ground in each side of the CB. These capacitors will influence the transients and reduce the differences between the waveforms.

4.10.5 Summary

We have seen how the bonding configuration, the grounding resistances and the existence of a grounding grid influence the both the current and voltage during the fault and after the fault. It was shown that because of mutual coupling the voltage in the sound phases increases during the fault and maybe also after the fault, depending on the value of the grounding resistances or the presence of a grounding grid.

Typically the larger the grounding resistance the larger is the voltage transient and the overvoltage. By other hand, the magnitude of the short-circuit current increases when the grounding resistance decreases, being maximum for the case where a grounding grid is used.

If the cable is connected to a shunt reactor the voltage in the sound phases after the disconnection is a decaying AC wave. Depending on the parameters of the circuit, the overvoltages may be larger than for an equivalent cable without shunt reactor. The overvoltage increase is larger if the shunt reactor is installed at the end with higher voltage magnitude at the disconnection instant. Thus, the magnitude is larger if the shunt reactor is at the end that is disconnected first.

We have also seen how a transformer affects the waveforms and how the type of connection changes the waveforms. A parameter that should always be modelled when simulating a short-circuit is the capacitance to the ground of the transformer as it will lead to a shaper TRV and increase the probability of a reignition/restrike.

Only SPGF were studied, however, all the theoretical principles were laid down and the readers can use them to analyse different types of faults and how the location of the fault affects the results. It is also advisable to download the example available online and study them following the available tutorials as they allow seeing the waveforms in different points of the cable and how the different parameters influence the waveforms.

References and Further Reading

1. IEC 60071-2 (1996) Insulation co-ordination—Part 2: application guide
2. IEE 60071-4 (2004) Insulation co-ordination—Part 4: computational guide of insulation co-ordination and modelling of electrical networks
3. Da Silva FMF (2011) Analysis and simulation of electromagnetic transients in HVAC cable transmission grids. PhD Thesis, Aalborg University, Denmark
4. Ibrahim AL, Dommel HW (2005) A knowledge base for switching surge transients. International Conference on Power Systems Transients (IPST), Canada, Paper No. 50
5. Alexander RW, Dufournet D (2008) Transient recovery voltage (TRV) for high-voltage circuit breakers. IEEE Tutorial: Design and Application of Power Circuit Breakers. IEEE-PES General Meeting
6. Liljestrand L, Sannino A, Breder H, Thorburn S (2008) Transients in collection grids of large offshore wind parks. Wind Energy 11(1):45–61
7. Ferracci P (1998) La Ferroresonance. Cahier technique n° 190, Groupe Schneider
8. Greenwood Allan (1991) Electrical transients in power systems, 2nd edn. Wiley, New York
9. Van der Sluis L (2001) Transients in power systems. Wiley, New York
10. IEC 60056-1987-03 (1987) High-voltage alternating-current circuit-breakers
11. IEEE guide for the application of sheath-bonding methods for single-conductor cables and calculation of induced voltages and currents in cable sheaths, IEEE Std. 575 (1988)
12. IEEE application guide for capacitance current switching for AC high-voltage circuit breakers, IEEE Std. C37.012 (2005)
13. IEEE Guide for the protection of shunt capacitor banks, IEEE Std. C37.99 (2000)
14. Cigre Joint Working Group 21/33 (2001) Insulation co-ordination for HV AC underground cable system. Cigre, Paris
15. Cigre Working Group B1.18 (2005) Special bonding of high voltage power cables. Cigre, Paris
16. Cigre Working Group C4–502 (2013) Power system technical performance issues related to the application of long HVAC cables. Cigre, Paris

Chapter 5
System Modelling and Harmonics

5.1 Introduction

In the previous chapter, we had the opportunity to analyse and study several electromagnetic transient phenomena. However, most of the examples were given for simple systems, consisting only of a cable, a voltage source and a transformer. Real systems are much more complex containing several lines, transformers, shunt reactors, generators, etc.

In this chapter, we will explain how to decide the modelling depth when studying different phenomena. To do that, we will use part of the West Denmark high voltage grid as planned for 2030.[1]

This network has several characteristics that make it ideal for the explanation given in the following sections. In 2030, the West Danish transmission network will be fully undergrounded at the 150 kV voltage level, but OHLs will still be used at the 400 kV voltage level. According to the existing grid plan, the network will comprise a total of 114 165 kV-cables covering a total of 2627.3 km, 27 400 kV-OHL covering a total of 1215.4 km and 36 165/400 kV-transformers. Figure 5.1 shows the network single-line diagram.

5.2 Modelling Depth for Switching Studies

5.2.1 Theoretical Background

The modelling depth, i.e. the number of busbars behind the node of interest that have to be included in the simulation model, has a strong influence on the simulation results. On one hand, if the models do not have enough detail the results are not accurate, while on the other hand if they have too much detail, the simulations will take too much time to run. Thus, a minimisation of the number of lines and busbars in the model while keeping the accuracy would represent an enormous

F. F. da Silva and C. L. Bak, *Electromagnetic Transients in Power Cables*,
Power Systems, DOI: 10.1007/978-1-4471-5236-1_5,
© Springer-Verlag London 2013

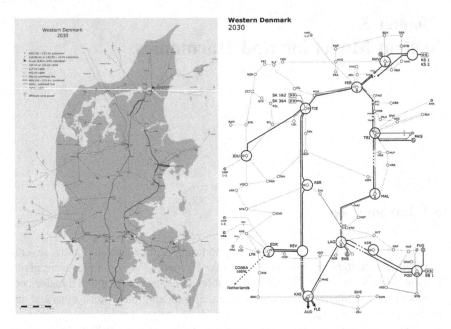

Fig. 5.1 West Denmark network as planned for 2030 (*source*: Energinet.dk)

time saving, which also generates financial savings. However, in order to minimise the model we have first need to know what we are studying, and second, how to do the simplifications.

We start by studying a method for the estimation of the required modelling depth when studying switching overvoltages or restrikes. In the study, we assume the worst-case scenario and that the voltage at CB at the switching instant is maximum. If the voltage at CB terminals at the switching instant is zero, it is unnecessary to model the network as no transient waves are generated, unless we are studying resonances, which we will learn how to do in Sect. 5.3.2.

We have seen in Sect. 4.3 that the peak voltage associated with cable energisation/re-energisation is not attained in the energisation/re-energisation instant, but some hundreds of micro-seconds later.

For an isolated cable, i.e. connected to an ideal voltage source, switched on at peak voltage, the instant of maximum voltage magnitude is normally a function of the velocity of the coaxial mode in the cable and it corresponds to the instant when the wave generated at switching reaches the receiving end of the cable for a second time. For a cable incorporated into a network, the peak voltage will normally occur at the same instant.[2]

Figure 5.2 shows the voltage in one phase during a restrike for different modelling depths. The CB is forced to restrike half-cycle after the switch-off when

---

[2] The exceptions are explained in Sects. 5.2.4 and 5.2.5.

Fig. 5.2 Voltage transient in one of the phases after the restrike for the different modelling depths. 5-busbars depth: *solid line*; 4-busbars depth: *dashed-line*; 3-busbars depth: *pointed line*; 2-busbars depth: *dashed-pointed line*

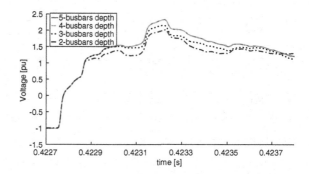

the voltage difference at the CB terminals is maximum and equal to 2 pu. For all four cases, it is used an N-1 equivalent network representing the grid around the respective modelled area.[3]

The voltage waveforms for a normal energisation would be similar to the ones of Fig. 5.2, but with lower oscillations due to the lower initial voltage difference at the CB terminals. The important result for us is that the peak voltage occurs at the same instant for both phenomena and modelling depths.

The peak voltage instant is normally independent of both the modelling depth and the energisation phenomena, i.e. normal energisation or restrike. On the other hand, the magnitude of the peak voltage is very dependent on both the type of phenomenon (because of the initial voltage difference at the CB terminals in the initial instant) and the modelling depth. The next step is to understand why the modelling depth affects the value of the peak voltage.

Two or more waves are generated when a cable is energised/re-energised. One of the waves propagates into the line being energised, whereas the other wave(s) propagate into the line(s) that are connected to the sending end of the cable being energised. The magnitude of each of these waves depends on the lines' respective surge impedances.

The waves propagating into the neighbour lines are partly reflected at the end of those lines and at the bonding points, if cross-bonded. These reflected waves are then refracted into the cable influencing the waveforms.

Figure 5.3 shows a simple example on how the waves act during the electromagnetic transient. The wave(s) injected into the neighbour line(s) (solid line in *I*) is reflected back at the line(s) end(s) (*II*) and is then refracted into the cable being energised (*III–IV*).

In order for the reflections of the neighbour busbars to reach the receiving end of the energised cable (*V*) before the peak overvoltage, the busbars have to be at a distance inferior to half of the distance travelled by the wave between the switch-on and the peak voltage instant. If the wave(s) reaches the cable receiving end

---

[3] For more information on equivalent networks see Sect. 5.2.6.

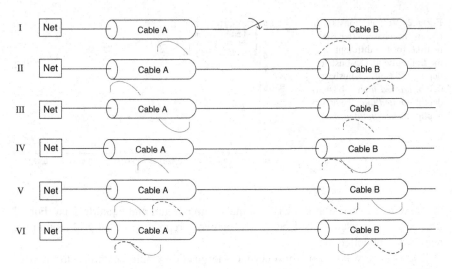

Fig. 5.3 Reflections and refractions of the waves generated at the switch-on instant. *Solid Line* Wave propagating into the adjacent line; *Dashed Line* Wave propagating into the cable being energised

before the expected peak instant (*VI*), changes will occur in both the waveform and the peak voltage.

Going back to Fig. 5.2, we can see that the 2-busbars waveform diverges from the other at approximately 0.4229s. This model has an equivalent strong network after 2 busbars, which is equivalent to a high reflection coefficient with a negative signal. Thus, in the 2-busbars model, the waves are almost fully reflected back with opposite polarity, whereas in the other models, they are only partially reflected back. As a result, the voltage becomes lower at 0.4229s for the model with a 2-busbars depth. The same happens for the 3-busbars depth model at approximately 0.423s, however, it only happens after the peak voltages instant for the 4-busbars depth model. Therefore, it can be concluded that for this particular example, it would be necessary to have a 4-busbars depth model in order to have an accurate result. A result that is particularly relevant since the existing IEC standard 60071-4 recommends a 2-busbars depth model as being accurate for the majority of situations.[4]

The reflections in the neighbouring cables can also affect the moment at which the peak voltage occurs even if such situations normally do not happen in a grid operating under normal conditions. Yet, there are two cases where this may not be true:

- A reflected wave reaches the cable precisely after the expected peak (either a coaxial mode wave or an intersheath mode wave);

---

[4] We do not intend to say that the standard is wrong, just that it is necessary to be careful for some configurations and be always critic of the results that are obtained in the simulations.

Table 5.1 Distance travelled by the wave during 470 μs for different wave speeds

Wave speed (m/μs)	Distance (km)	Busbar distance (km)
300 (light speed)	141	70.5
180 (coaxial mode)	85	42.5
80 (intersheath mode)	38	19

- An OHL is installed in the vicinity of the cable;

Both situations are analysed later in this chapter.

It should also be noticed that the estimation method we are going to learn does not work for weak grids or the simulation of black-start operations, where it is typical to model most of the network due to the reduced number of lines.

5.2.2 Calculation of the Modelling Depth

The previous section explained how the peak voltage depends on the modelling depth, and how important it is to know which lines/busbars to model. This section presents a method which can be used to obtain that information. The method is first explained through a practical demonstration for the energisation/re-energisation of a cable installed in the northeast of Denmark between the NVV-BDK busbars (see Fig. 5.1). The same cable was used in the example shown in Fig. 5.2.

For the energisation/re-energisation of this specific cable, the peak voltage is attained 470 μs after the wave has reached the cable receiving end (see Fig. 5.2).

Table 5.1 shows, the distances travelled by a wave during 470 μs for different typical wave speeds.[5] In order for the reflections of the neighbouring busbars to reach the receiving end of the NVV-BDK cable before the peak overvoltage, the busbars have to be at a distance lower than half of the distance travelled by the wave in that period of time as the wave has to reach the reflection point and be reflected back to the cable receiving end.

Different approaches can be applied. The worst-case scenario is to consider that the wave speed is equal to the speed of light and to model all the lines up to a distance of 70.5 km from the NVV busbar. This is an incorrect approach because in a cable, the wave speed is approximately half the speed of light. Therefore, such an approach would result in excessive modelling and longer simulation times.

The situation changes when OHLs are in the vicinity of the cable being (re)energised. In that case, the wave in the OHL travels at a speed close to the speed of light. In this example, the 165 kV network is a pure cable network, and the wave speed is the cable's coaxial mode speed.

So, by knowing the grid configuration, the wave speed in the different lines of the system and the time required to have the first peak, the required modelling

---

[5] The speed of the coaxial mode and intersheath mode shown in the table are slightly larger than the usual ones for reducing the risks of an incorrect estimation of the modelling depth.

depth can be estimated. As an example, if the cables all have the exact same characteristics, i.e. the wave velocity is the same for all, we know that it is necessary to model all the cables up to busbars that are a distance equal or lower than the length of the cable being energised.

The next step is to know how to automatise the process so that we do not need to hand calculate the required distance for all cases.

Typically, utilities have a PSS/E file, or equivalent, containing the entire grid. This file contains information on all grid equipment (lines, shunt reactors, transformers, generators, loads, etc.), respective characteristics and connections between busbars.

This file can be used to obtain the grid layout and to design two matrices equivalent to the grid, one containing the distance between the busbars and the other the wave speed in each line.

Figure 5.4 shows part of the network in the vicinity of the NVV busbar, whose equivalent distance matrix is shown in (5.1). The wave speed matrix is equivalent to the matrix shown in (5.1), but containing the wave speed of each line instead of the length. The distance between the busbars can typically be obtained directly from the files describing the network, but the same is not possible for the wave speeds, since these files are normally used for steady-state studies and lack information about the lines geometry. One solution could be to use values that are slightly larger than the average wave speed values, e.g. 180 and 280 m/µs for the coaxial speeds of a cable and an OHL, respectively, or to create an extra file containing the precise wave speed in each line for the different lines. Another possibility is to approximate the wave velocity using the lumped parameters, however, only an approximation of the coaxial mode speed can be obtained using this method.

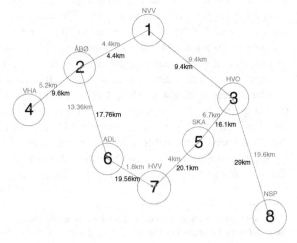

Fig. 5.4 Busbars in the vicinity of the restriked node. *Red* Distance between nodes; *Blue* Distance between the nodes and the NVV node

In general, it is only necessary to model the elements whose voltage level is the same as that of the cable being switched. Thus, the matrices and the designed model need to contain only busbars with the same nominal voltage.[6]

$$
\begin{array}{c c c c c c c c c}
 & B1 & B2 & B3 & B4 & B5 & B6 & B7 & B8 \\
B1 & 0 & 4.4 & 9.4 & 0 & 0 & 0 & 0 & 0 \\
B2 & 4.4 & 0 & 0 & 5.2 & 0 & 13.36 & 0 & 0 \\
B3 & 9.4 & 0 & 0 & 0 & 6.7 & 0 & 0 & 19.6 \\
B4 & 0 & 5.2 & 0 & 0 & 0 & 0 & 0 & 0 \\
B5 & 0 & 0 & 6.7 & 0 & 0 & 0 & 4 & 0 \\
B6 & 0 & 13.36 & 0 & 0 & 0 & 0 & 1.8 & 0 \\
B7 & 0 & 0 & 0 & 0 & 4 & 1.8 & 0 & 0 \\
B8 & 0 & 0 & 19.6 & 0 & 0 & 0 & 0 & 0
\end{array}
\tag{5.1}
$$

Using the matrix describing the system, we can write a code to search in the matrix all the possible paths up to a certain depth. For safety, we should choose a depth larger than we expect (e.g. 7–8 busbars is usually a safe number for a HV network).

Then we calculate for all the paths which busbars are located up to a distance of less than 0.525 times the peak time of the cable being energised/re-energised.[7]

When applying this method to the energisation of the NVV-BDK cable, we know that we have to include all the cable up to a distance of 67 km, which is the distance equivalent to 712.5 µs for a coaxial speed equal to 180 m/µs. Using the method it is estimated 4-busbars as the minimum modelling depth, which we have seen in Fig. 5.2 to be correct.

The estimation method provides not only the required modelling depth, but also the exact busbars that have to be included in the model. Using this information, the model can be optimised, and instead of having a modelling depth of 4-busbars, it can have only the necessary busbars.

According to the estimation method, the simulation of the NVV-BDK cable switch-on from the NVV side, needs 14 busbars, whereas the use of a 4-busbars modelling depth corresponds to a total of 29 busbars. Thus, the optimisation of the model to only 14 busbars would represent a substantial reduction of the simulation time. Figure 5.5 compares the voltage in the cable receiving end, similar to Fig. 5.2, for a model with a 4-busbars modelling depth (29 busbars) and another model with only the required 14 busbars. The voltage peak magnitude is the same for both models, and the divergences between the models only appear after the peak instant.

---

[6] However, a transformer as well as the busbar in the other side of the transformer should be included in the model, when a transformer is inside or in the boundary of the area being modelled.

[7] Ideally it would be 0.5, but it is given a security margin for the case of a reflection reaching the cable immediately after the expected peak time, which can increase the voltage magnitude.

Fig. 5.5 Voltage transient in one of the phases after the restrike for a 4 busbar modelling depth (*solid line*) and a model containing only the required busbars (*dashed line*)

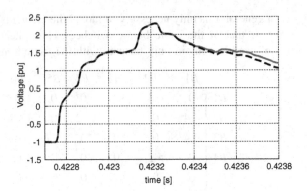

5.2.3 Modelling of the Cross-Bonded Sections

We know by now how to estimate the number of lines that should be included in the model. However, if these lines are cross-bonded cables they may have several minor sections. Thus, it would also be helpful to know if it is possible to minimise the number of minor sections in the model reducing the simulation time even more.

An accurate modelling of all cross-bonded sections is indeed necessary for the simulation of some electromagnetic transient phenomena, as the wave during a transient can be reflected at the cross-bonded points. We saw in Sects. 4.3 and 4.8.1 how the reflections at the transpositions points, influence the waveform and the peak voltages for cables connected to an ideal voltage source, especially because of the grounding of the major-sections. Whereas the need to model all cross-bonding sections in the cable being energised is clear, the same may not be true for the neighbour cables.

We have seen in Sect. 4.3.4 that the coaxial mode voltage does not influence the voltage in the screens, whereas the intersheath mode voltage, which depends on the screens voltages, influences the voltage in the conductor. Thus, an accurate modelling of the screens is only required if the intersheath mode has enough time to reach those screens and return to the cable being energised prior to the voltage peak instant, similar to what was done for the coaxial mode.[8]

We have also seen that the propagation speed of the intersheath mode is lower than the propagation speed of the coaxial mode. As a result, in some situations not all the cables included in the simulation model need to have a detailed model of all major-sections and instead the number of major-sections is reduced to one or even changed to both ends bonding.

In order to estimate which cross-bonded sections need to be accurately modelled, we use the same method that was used to calculate the modelling depth,

---

[8] This assuming that we are interested in an accurate simulation of the voltage. The same is not true for the current where it would be necessary to consider the intersheath mode currents generated by the coaxial current.

but substituting the coaxial mode speed of each cable by the intersheath mode speed, which can be approximate with a security margin to 80 m/μs.[9,10]

For the particular example of a restrike in the NVV-BDK cable, it would be necessary to model all the cross-bonded sections of the system due to fact that all the cables installed near the NVV node are short. Thus, the theory is demonstrated using another area of the network where it can be demonstrated that it is not necessary to model all the bonding sections of the more distant cables.

The method is exemplified to a restrike in the STS end of the STS-LKR cable installed in the west part of Denmark.

The application of the method previously described indicates that five busbars have to be included in the model. As the ASR busbar is connected to two transformers, a sixth busbar is added, so that the transformers are included in the model and the reflection in that busbar is properly represented. Figure 5.6 shows the modelled single-line diagram for the system behind the energised cable.

According to the method it is necessary to have accurate modelling of the cross-bonded sections in only two of the cables, more precisely; the two cables adjacent to the restriked cable: the STS-GNO and STS-ASR cables (see Fig. 5.6).

Example
To simplify the analysis and comparison, the system is kept as simple as possible and it is considered that all cables of the system have only two major cross-bonded sections.

Three different models are prepared for comparison:

Fig. 5.6 Single-line diagram of the simulated system. *Dashed lines* Cables that need an accurate modelling of the cross-bonded sections

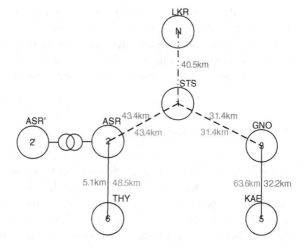

---

[9] It will usually be less than 80 m/μs (for example, it was around 60 m/μs for the example shown in Sect. 3.4.2).

[10] We should remember that we can use the precise intersheath mode velocity of each cable if we have enough information available.

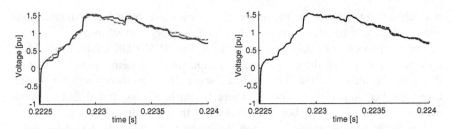

Fig. 5.7 Voltage in the cable receiving end during re-energisation. **a** *Dashed line* Model 1; *Solid line* Model 2; **b** *Solid line* Model 2; *Dashed line* Model 3

Model 1: The switched on cable model has two major-sections and all the remaining cables are modelled with one major-section;
Model 2: Equal to Model 1, but the two cables adjacent to the STS node (STS-GNO and STS-ASR) are modelled with two major-sections, whereas the remaining cables with one major-section;
Model 3: Equal to Model 2, but with two more cables (GNO-KAE and ASR-THY) modelled with two major-sections;

According to the theory presented, Model 1 has an insufficient level of detail, Model 2 has the minimum level of detail required to an accurate simulation of the maximum voltage peak and Model 3 has an excessive level of detail.

Figure 5.7a shows the voltage in the receiving end for the re-energisation of the STS-LKR cable using Model 1 and Model 2. Figure 5.7b shows the same results for Model 2 and Model 3.

The results confirm the accuracy of the method. The differences between Model 1 and Model 2 are visible before the peak, while the differences between Model 2 and Model 3 are only visible after the peak instant. Thus, it is not necessary to make a detailed modelling of all the major-sections for the two outermost cables, in order to obtain a precise value of the maximum peak voltage.

5.2.4 Possible Inaccuracies

The presented method assumes that the peak voltage occurs when the transient wave reaches the receiving end of the cable by the second time and also that this instant is independent of the network. Something that is not always true, either because of a combination of low and/or large reflections coefficients in the network, or because of a magnification effect resultant from different surge impedances. Possible examples are systems with:

- Very few lines;
- Low reflection coefficients in the joint points, meaning that the cables are connected in series and there is almost no load/generation in those points;

- An OHL installed in the cable vicinity;

The first two cases are explained below, whereas the last is explained in the next section.

In the demonstration of the inaccuracy, we use a system consisting of only two cables, *Cable A* and *Cable B*, both bonded at both ends. The cables' characteristics are the same, i.e. the surge impedance is the same for both, and they are connected in series with *Cable A* connected to an ideal voltage source. The inaccuracy of the method presented in the previous section is demonstrated for an energisation of *Cable B*.

Cable B is energised and one wave propagates into it, while another wave propagates into *Cable A*. The wave propagated into *Cable B* is reflected back at the cable receiving end and it eventually reaches the cable joint point. Normally, part of the wave would now be reflected back into *Cable B* and the transient peak voltage would be at the moment that the reflected wave reaches the cable receiving end.

However, the cables have equal surge impedances and nothing else is connected at the joint point. Thus, there is virtually no wave reflection in the joint point,[11] and the transient wave is entirely refracted into *Cable A*. Such a situation does not normally occur in a normal system where part of the wave is reflected back into the cable because of the number of lines and transformers also connected to the node.

The wave propagates then in *Cable A* reaching its sending end where it is reflected back with opposite polarity,[12] and it will eventually arrive at the receiving end of *Cable B*. The peak voltage will occur at this instant, i.e. after the moment initially expected for the peak voltage.

Figure 5.8 shows the waveform during the transient for the situation just described, where it can be seen that the peak instant occurs later than expected.

An example of a system under these conditions is the study of a network for a black-start operation. In this case, the generation is reduced and many loads are

Fig. 5.8 Example of the voltage in the end of Cable B (*dashed line*) and the joint point (*solid line*) during the energisation of Cable B

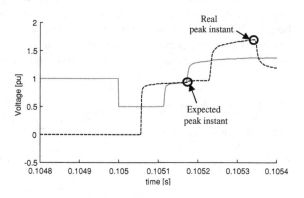

---

[11] There is a small reflection because of the grounding of the cable's screen.

[12] We are considering the presence of an ideal voltage source in this point.

disconnected from the system allowing a replication of the conditions previously described.

If the peak overvoltage comes after the expected instant, it becomes necessary to estimate the modelling depth for the new peak voltage instant again. Thus, in our simulations, we always need to confirm that the peak voltage is at the expected instant.

However, sometimes, the overvoltages after the expected instant may not be real, but rather a result of simplifications in the model.

Figure 5.9 shows a network reduction. To maintain the simplicity, the network is reduced from one busbar depth to one cable. The *Eq. B* represents the lower/higher voltage levels connected to the node. The *Eq. A* represents the entire equivalent network seen from the sending end of Cable A.

The *Eq. C* is the series of *Eq. A* with *Cable A* in parallel with *Eq. B*. Thus, the impedance of *Eq. C* is always lower than the impedance of *Eq. A*.

The smaller the impedance of the equivalent network, the larger the reflection coefficient and consequently also the reflected wave. Therefore, a model with fewer nodes will have a larger reflection coefficient in the boundary nodes and is more likely to show an overvoltage after the expected instant. In addition, the more nodes a simulation model has, the later is the reflection in a boundary node and the more damped is the wave before the reflection reaches the receiving end of the cable being energised/re-energised.

Thus, it is possible that the late overvoltage seen in the simulations using a simple model is not seen when using a more complex model. Figure 5.10 shows an example of such a case; the simulation is for the re-energisation of the same cable, but while the simple model shows a larger overvoltage at approximately 0.225s, the complex model does not show such overvoltage, which allows us to conclude that the overvoltage is not really present and is resultant from an oversimplification of the model.

It is important to notice that in principle, if one does not see an overvoltage after the expected instant in the model calculated according to the method previously explained, it should neither be present in a more complex model, i.e. with more busbars, as the reflection coefficients are smaller. Thus, the method previously explained provides accurate results as one can detect the cases where the overvoltage is after the expected instant. In these cases, the estimation method should

Fig. 5.9 Possible network reduction, where *Eq. C* is equal to the parallel of *Eq. B* with the series of *Cable A* and *Eq. A*

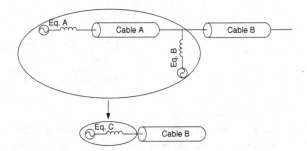

Fig. 5.10 Voltage in the cable receiving end during the de-energisation. *Dashed line* Simple model; *Solid line* Complex model

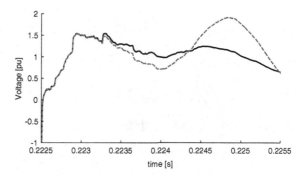

be run again, but using the new peak time. It should be noted that this overvoltage may disappear when the simulation is run in the more complex model.

There may be, however, some special configurations; more precisely, areas with many short cables with different electric characteristics that may lead to a voltage build up, where the previous explanation may not be applied. Yet, such configurations should be unusual in a transmission network.

5.2.5 Extended Method and Minimisation of the Inaccuracies

The inaccuracy explained in the previous section is a result of the reflections in the outmost nodes of the model. One can minimise this problem by a combination of FD-models and lumped-parameters models. It is true that the addition of the lumped-parameter models increases both the system complexity and the simulation running time, but much less than if FD-models were used all over.

Typically the simulation of a FD-model is usually at least ten times longer than the simulation of an equivalent lumped-parameter model, plus the extra time necessary to design the FD-model. A simulation model can easily have several dozens of FD-models, when cross-bonded cables are present. In this situation, the adding of lumped-parameter models will have an almost negligible effect in the total simulation time.

Like before, the cables adjacent to the energised/re-energised cable are modelled with maximum detail, i.e. FD-models and all the minor-section modelled with the exact lengths, as the distance increases the cables are still modelled by means of FD-models, but with only one equivalent major cross-section or ideal cross-bonding. According to the method previous explained, the model would now be complete.

To avoid the high reflections that could lead to inaccurate results after the expected peak and possible mislead us when running the simulation, it is added a third level where the cables/lines are modelled by means of lumped-parameters

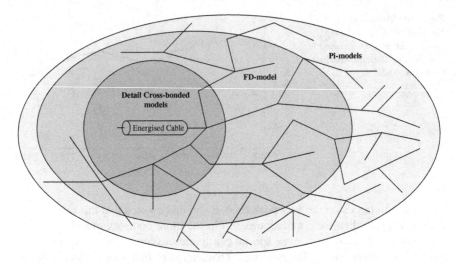

Fig. 5.11 Diagram of the final estimation method with the three levels of detail

models. This third level increases the simulation running time, but considerably less than if FD-models were used, and have a small influence in the total running time. Finally, an equivalent network is used to represent the remaining network like done before.

Figure 5.11 shows a diagram with the three levels, where the solid lines represent cables and OHLs.

The question is how to define the depth of this third level made of lumped-parameters.

The more efficient approach is to model from the beginning an extra level, i.e. one 1-busbar depth, by means of lumped-parameters. This approach increases the trust of the user in the results at the expenses of only a small increase of the simulation running time.

It is then verified if the peak overvoltage is after the expected instant. If it is, it is added another level to model and the process is repeated. If one wants to be completely secure, it can be modelled the remaining network by means of lumped-parameters, but this high level of detail is normally unnecessary.

Thus, in summary the model should have a first zone, closest to the switched cable, with a full modelling detail, i.e. FD-models are used and all cross-bonded sections are included in the model. A second zone, further out still with FD-models, but with the cross-bonding simplified to one major cross-bonded section. A third zone, with one busbar depth and lumped-parameter models. This reduces the likelihood of erroneous high overvoltage at the expense of a relatively small increase in the computational effort. A 2-port or N-port boundary equivalent network represents the system outside these three zones.

5.2.6 Equivalent Network

We have talked about the use of equivalent networks several times in the previous sections, but we have not yet described them.

One of the most common methods used for design equivalent networks is the 1-port equivalent network. This method consists in connecting a voltage source in series with a RL impedance, calculated for the steady-state frequency of the short-circuit impedance at each boundary bus of the network. Sometimes there is added a shunt load with the surge impedance of the line attached to the bus, which improves the high-frequency response of the equivalent network, but at the cost of a worse low-frequency response.

Another possibility is the N-port equivalent network. In this method the sources of the 1-port network are interconnected by means of lumped-parameters lines, providing a more accurate representation of the system.

In any case, the impedance of the equivalent network at the boundary nodes is much lower than the impedance of the lines that are in reality connected to those nodes. As a result, the magnitude of the reflected wave is larger than in reality, resulting in possible misleading results. Thus, there is necessity in having the equivalent network at a certain distance from the line being energised.

Another important parameter of an equivalent network and its distance to the node of interest is how it affects the frequency response. We will deal with this topic in Sect. 5.3.2.

5.2.7 Systems with Cables and OHL

By this point, we have a method that we can use to estimate the required modelling depth, but we have not yet made any reference to cases where OHL are installed in the vicinity of the area being modelled. This is important at the presence of an OHL represents an extra difficulty when estimating the modelling depth because of the reflections and refractions in the cable-OHL junctions.

Three different scenarios are possible:

- Scenario 1: The OHL(s) is/are inside the area to be modelled;
- Scenario 2: The OHL(s) is/are adjacent to the modelling area, i.e. the OHL is the first line not to be included in the model;
- Scenario 3: The OHL(s) is/are distant from the modelling area;

Scenario 1 is not a problem as the OHL(s) would always be included in the model. We just have to be careful in considering that the wave speed is higher in an OHL than in a cable.

When facing Scenario 2, it is necessary to include the OHL in the model as well as the cable(s) immediately adjacent to the OHL, as a voltage wave is magnified when flowing from a cable into an OHL and its reflection later may change the magnitude of the peak voltage.

Scenario 3 is a little more complicated, however, the method described in the previous section can still be apply as long as we analyse the model and the results with criteria.

The reason of concern is the voltage magnification when the wave flows from the cable into the OHL. When we do the first estimation of the model we can see our far are the boundaries of our model from the OHL(s). If it is a long distance we can expect that the OHL(s) will not influence the results,[13] and thus we do not need to include it in the model. If the OHL(s) is not installed very far we extend our model in order to included, but using lumped-parameters, so that we do not increase the simulation running time.

The lumped-parameters models contain information of the surge impedances of the lines and show the voltage magnification. Moreover, a larger voltage variation is normally expected when using lumped-parameters models. Not only is there more damping in a FD-model than in an equivalent lumped-parameters model, because of skin and proximity effects, the voltage oscillation is also larger in the latter, because of the presence of lumped capacitors and inductors.

Thus, when we run the simulation using this model we can see if there is, or there is not, an overvoltage after the expected peak instant. However, as the use of lumped-parameters aggravates the overvoltage we need to replace the lumped-parameter models by FD-models and run the simulation again in order to verify if the overvoltage is real or a result of inaccuracy in the simplified model.

5.3 Harmonics in Cable-Based Systems

5.3.1 Introduction

A frequency spectrum calculated for a given point of a cable-based network is substantially different from a frequency spectrum calculated for the same point of an equivalent OHL-based network.

First, the cable network has its resonance points at lower frequencies. As a result, undesirable resonance phenomena are more likely to occur in a cable-based network.

Second, the magnitudes of the higher frequencies are lower in the cable. Consequently, the higher frequencies are less damped in a cable-based network.

Thus, it is important the understanding of these differences when working with cables. Moreover, a frequency spectrum can also provide information about what transients to expect and it is a valuable tool that can be used for insulation co-ordination studies.

---

[13] The term long distance is seen in comparison with the size of the line being energised. For a 1 km long line, 10 km is a long distance, but the same is not true for a 50 km long line.

5.3.2 Estimation of the Frequency Spectrum

We have studied the frequency spectrum for a standalone cable in Sect. 3.5, but normally we want to know the frequency spectrum for a point of the network, in order to verify if there are any harmonic related problems or resonances at undesired frequencies.

However, unless frequency scan field measurements are available, it is not feasible to do an accurate FD-network without designing a large area of the network by means of FD-models. Thus, simplifications are necessary.

The normal approach is to divide the system into two areas, a detailed study area and an external network area. As an example, the method we saw in Sect. 5.2 is a variation of this classic principle.

The conventional approach used for the modelling of the external network is to match the frequency response of the original network with a lumped-parameter network. Besides not being completely exact, this approach has the problem that it is necessary to know the frequency response of the system, seen from the analysed area. This can, if necessary, be done for a small system, but not for a large system.

As we have seen several times before, the size of the detailed area, or in other words the modelling depth, influences the simulation results. We have also seen that the type of bonding used in a cable affects its frequency spectrum. Section 5.2 provided guidelines for the design of the detailed network when simulating switching overvoltages, but not for resonance phenomena, which takes us to out next question. How much detail do we need to put in our models when estimating the frequency spectrum for a given busbar of the network?

The common approach in this case is to:

- Design a detailed system up to a distance of two or three busbars from the point of interest and use an equivalent network for the rest of the grid;
- Repeat the previous point, but increase the modelling depth of the detailed area in one busbar;
- Compare the frequency spectrums for both systems;
- Repeat the process until the difference between the spectrums is minimum around the frequencies of interest;

Similar to Sect. 5.2 different modelling approaches can be used:

- An equivalent network is designed: The detailed area is modelled by means of FD-models. The external network is modelled by means of an equivalent network (this approach will be named *Di-Eq*, where *i* is the modelling depth of the detailed area);
- The entire system is modelled, or at least a large part of it: The detailed area using FD-models and the external network by means of lumped parameter, for

the lower voltage levels an N-port equivalent network is used,[14] (this approach will be named *Di-L*, where *i* is the modelling depth of the detailed area);

Besides comparing these two modelling approaches the influence of the number of major-sections in the frequency spectrum, when using cross-bonded cables, should also be taken into account, as its simplification may represent an important reduction in both simulation time and model complexity.

Figure 5.12 compares the frequency spectrums for the two modelling approaches previously described; using a N-ports equivalent network for the *Di-Eq* modelling approach. The cables modelled by means of FD-models are the same in both models, the difference lies in the modelling of the remaining network. The simulated network is the West Denmark transmission network as planned to 2030.

The frequency spectrum is obtained for a point in the 150 kV network, where a 22.1 km cross-bonded cable is connected to a phase shift-transformer (LEM node), the same used in the examples shown in Fig. 3.22, resulting in a low-frequency resonant point. The cable is open in the receiving end so that the frequency scan indicates the frequencies excited during the energisation transient (more on that in Sect. 5.3.3).

The observation of the frequency spectrums indicates that:

- The number of resonance frequencies of the L-Net is equal or superior to the number of resonance frequencies of the equivalent Eq-Net;

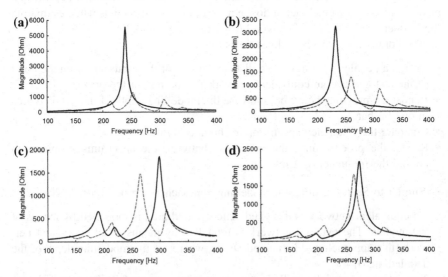

Fig. 5.12 Frequency spectrums for different modelling approaches: *Dashed Line* Model Di-L; *Solid Line* Model Di-Eq; **a** Model D1; **b** Model D2; **c** Model D4; **d** Model D5

---

[14] As an example, if we are modelling a transmission network we are not going to model also the several distribution networks, we will instead use an equivalent network.

- The magnitude of the impedance at the main resonance frequency (~ 250 Hz) is larger in the Eq-Net than in the equivalent L-Net;
- The modelling depth has more influence on the Eq-Net frequency spectrum than on the L-Net frequency spectrum;
- Only the model with a 5 busbars modelling depth presents similar results for both modelling approaches.

An Eq-Net is always less complex than the corresponding L-Net, with the notable exception of the limit case when the entire network is modelled. As the modelling depth increases, the number of lines and other elements in the model also increase, followed by an increase in the size of the N-port external equivalent network. As a result, an increasing number of resonance points are expected.

The L-Net contains all the network generators, transformers and loads, and the only difference when the modelling depth increases lies in the modelling of some of the cables, which changes from lumped-parameters models to FD-models. Thus, an increase in the modelling depth does not affect the frequency spectrum of the L-Net as much as in the Eq-Net. A corollary of the previous paragraph is that the Eq-Net model may require the modelling of a large area of the network in order to present accurate results.

The simulation is expected to be slower for the L-Net approach than for the equivalent Eq-Net model. However, the Eq-Net requires more cables modelled by means of FD-models in order to provide accurate results. The simulation of a FD-modelled cable is usually ten times slower than the simulation of the equivalent lumped-parameters cable. Moreover, if the cable is cross-bonded, the simulation of the FD-modelled cable is then $10*x$ times slower, where x is the number of minor sections. Consequently, contrary to what common sense would initially indicate, an accurate simulation can very often be faster for a L-Net than for an Eq-Net.

Assuming that we have decided to use the L-Net, we still need to decide the size of the detailed network and of the external network (assuming that we are not going to model the entire network). There are no clear rules that can be given for these situations and an engineer needs to rely on his/her experience. The safest process it to continue using the approach previously explained of increasing the network size until the frequency spectrum stops changing. However, we know that it is not very time consuming to have more lumped-parameter models, because the majority of the simulation burden is in the FD-models. Thus, it is usually more sensible to model a large area using lumped-parameters models and to do the sensitivity analysis for a small area near of the point of interest using FD-models.

Summary
We have seen that the L-Net has a much larger accuracy than the Eq-Net, having also the advantage of being less sensitive to changes in the modelling depth. These conclusions were expected and are not surprising.

The difference is that we are doing frequency studies for cable-based networks, instead of the more usual OHL-based networks. The simulation of a cable by means of a FD-model is more time consuming than the simulation of an OHL. As a result, the relative increase in the simulation time when using a L-Net instead of an

Eq-Net is small for a detail area with several dozens of cables. As an example, we can see in Fig. 5.12 that the D2-L present more accurate results than the D5-Eq model while having a lower simulation time.

Consequently, it is advised to use the L-Net instead of an Eq-Net as it will yield more accurate results at the expense of a relative small increase in the simulation time.

The required modelling depth can be obtained using the classic method of increasing the number of cables modelled by means of FD-models until the frequency spectrum stops changing. After obtaining the frequency spectrum it is also possible to do some mathematical treatment of the result and design new models showing the same frequency spectrum, but with a lower computational burden.

5.3.3 Frequency Spectrums and Electromagnetic Transients

We know that many electromagnetic transients are associated to high frequencies. Thus, it would in some situations be helpful if we could predict the frequency of the electromagnetic transient from a frequency spectrum so that we could tune or model to that frequency range and also validate the simulation results.

In order to demonstrate the principles that we are dealing with, we focus our attention in the energisation of a cable. We have seen that the transient waveforms are mainly related with the length of the cable being energised, the lengths of the neighbour cables and the reflection/refraction coefficients. The frequency spectrum in a point of the network depends on the same factors, thus, we can may be use frequency spectrums to provide information on what to expect in an electromagnetic transient and to calculate the frequency of the transient.

As an example, if the length of the cable being energised increases, the frequency of the transient decreases as the distance travelled by the wave without being reflected is larger. Similarly, if the cable's length increases the capacitance and inductance also increase, whereas the resonance frequency decreases.

Methodology

We should obtain the frequency spectrum for the same conditions in which we are going to energise the cable. Thus, the cable being energised should be connected to the network at the sending end, but open at the receiving end. The frequency spectrum is obtained at the cable receiving end and the CB connecting the cable to the rest of the network is closed.

The frequencies of the transient correspond to the parallel resonance points of the frequency spectrum. The main frequency will typically be the first parallel resonance point, which has also normally the larger magnitude.

We should remember that the magnitudes of the different frequency components of the transient depend on the voltage at the energisation instant. As a result, the frequency spectrum does not give any indications on the magnitude of the transient and its transient, but only on the associated frequencies.

A final note on how to obtain the frequency of a transient wave. The transient has usually a short duration, thus, the use of the Fast Fourier Transform (FFT) does not provide accurate results. It is normally better to use a Short Time Fourier Transform or Wavelets.

5.3.4 Sensitivity Analysis

The existing IEC standards IEC-62067 and IEC-60840 allow a deviation in the thickness of some of the cable layers of up to 10 %.[15] These variations may have a strong influence in both the transient waveforms and the frequency spectrum, which we have seen to be related.

It is safe to say that the thicknesses of the screen and outer insulation of the cable do not have a large influence except for short-circuits. Thus, only the thicknesses of the cable conductor and insulation are changed and the frequency spectrum estimated.[16] The layers are analysed separately, meaning that for a deviation in the conductor thickness, only the conductor thickness is changed.

Figure 5.13 shows the frequency spectrum for a variation of 5 % in the insulation thicknesses when using the model D2-L (Sect. 5.3.2), similar variations would be seen for other modelling depths. The increase in the insulation thickness results in a decrease in the capacitance. Thus, the resonance frequency increases when the insulation thickness increases and decreases when the thickness decreases.

A change in the thickness of the cable insulation will result in larger changes at the ~ 250 Hz resonance frequency than at the other resonance frequencies. The ~ 250 Hz resonance frequency is the direct result of the cable-transformer interaction. Consequently, a change in the cable is more noticeable at this resonance frequency.

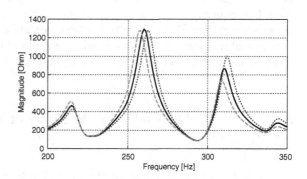

Fig. 5.13 Frequency spectrum for a variation of 5 % in the insulation thicknesses. *Solid line* Reference; *Dotted line* Thickness increases; *Dashed Lines* Thickness decreases

---

[15] Remember that these values may change in future iterations of the standard.

[16] We have to remember that changes in the thicknesses also correspond to changes in the resistivity and permittivity as seen in Sect. 1.1.

The behaviour of the impedance magnitude is not so linear. It increases, for example, when the thickness decreases for the first resonance point (~ 210 Hz), but it has the opposite behaviour for the main resonance point (~ 250 Hz).

The increase of the capacitance is normally followed by a decrease in the impedance at resonance frequency, e.g. a parallel LC circuit. The opposite behaviour at ~ 210 Hz is explained by a high capacitance and inductance of the network behind the transformer, including the transformer at that specific frequency.

This example shows that even for the standard tolerance the variation of the frequency spectrum can still be considerable and it may be the difference between exciting or not exciting a resonance frequency, i.e. between having or not having very high currents/voltages.

5.3.5 Conclusions

It was shown that the frequency spectrum and the transient harmonic content are correlated. Therefore, the adequate modelling depth for the simulation of a resonance can be based on the frequency spectrum as seen from the node to be simulated.

The classic method, which is used for most types of simulation, is to compare the frequency spectrum for increasing modelling depths and to stop when the frequency spectrum stops changing. However, the total simulation time of a cable-based network is mostly a function of the amount of cable modelled by means of FD-models. Consequently, the modelling of the network outside of the area of interest by means of lumped-parameters models does not represent a substantial increase in the total simulation time, whereas the accuracy of the frequency spectrum increases substantially.

The cable capacitance changes when the thickness of the conductor or insulation changes, being, however, scarcely affected by changes in the other parameters. Consequently, the frequency spectrum is affected if the datasheet values are incorrect or the permittivity and resistivity constants are not properly corrected.

The total frequency deviation depends on the model used, but the use of a more complex model does not result in a lower frequency deviation.

5.4 Type of Cable Model for the Study of Different Phenomena

The study of different phenomena requires different models. Although it is clear that the more accurate models are usually the FD-models, these models have the drawback of requiring more computer power and of being more difficult to prepare when compared with other models.

Therefore, it would be a big help to have guidelines to which type of model is more suitable for each phenomenon. These guidelines have been partially given before, but it is more helpful to have all them together in one place and that is the objective of this section.

These guidelines are not to be blindly followed, but to be used as a first indication when designing a simulation model. They are given for generic situations, but specific cases may require different models than those hereby proposed. The suggestions are also given with a certain security margin. As an example, it is suggested to use FD-models when simulating the switching overvoltage of a cable, however, a more experienced engineer may prefer to calculate the target frequency and use Bergeron models, which are faster are more stable.

Another important factor is the software that is being used. Different software use different mathematical methods for the modelling and fitting of the models, mainly for the FD-models, and for some cases it may be rather difficult to setup these models. The list given below uses FD-model as a generic term and it is up to the reader to decide if he/she is capable of validate these models.

The stability of the simulation needs to be also taken into account. The more complex the model the higher is the probability of numerical problems. Thus, it may be necessary to use simpler models when modelling big/complex systems.

Yet, in the end, it all resumes to one important question: *How accurate does the results needs to be?*

Switching overvoltages are typically below the maximum voltage limit. Thus, a simpler model can be used without problems, as a certain amount of inaccuracy is acceptable. Moreover, models without frequency dependence provide larger overvoltages than FD-models, because of the lower damping for the higher frequencies. Therefore, it is typical to do a first set of simulations using the simpler models and if the obtained results are close or above the defined threshold the simulations are redone using the more complex models.

The same reasoning can be applied to the modelling depth. We have seen that models with fewer nodes tend to give larger overvoltages. Thus, a simpler model may be used in the first set of simulations, being the simulations repeated using more complex models if the results are close or above the defined threshold.

For resonances the important is often the damping of the cable and surrounding equipment at the resonance frequency, thus, the important is to be certain on the frequency of the resonance point and to have an accurate modelling for that frequency.

Table 5.2 suggests for different electromagnetic phenomena the type of cable model to use, the detail that should be putted into the modelling of the bonding and the modelling depth.

Table 5.2 Cable models suggested for different types of simulations

Phenomenon	Model	Bonding detail	Modelling depth
Switching overvoltage	FD-models	See Sect. 5.2.3	Depends on the lengths: See Sect. 5.2.5
Zero-missing	Bergeron or lumped	Not required	Target cable
De-energisation	Bergeron or lumped	Not required	1 busbar
Restrikes	FD-models	See Sect. 5.2.3	Depends on the lengths: See Sect. 5.2.5
Series resonance	Mix of FD-models and lumped[1]	No[2]	See Sect. 5.3.2
Parallel resonance	Mix of FD-models and lumped[1]	No[2]	See Sect. 5.3.2
Ferroresonance	Mix of FD-models and lumped[1]	All	See Sect. 5.3.2
Faults	Bergeron	All of the faulted cable	1 busbar
Frequency scans and harmonic sources	Mix of FD-models and lumped	No[3]	See Sect. 5.3.2

[1] For the resonances is necessary to be aware that the damping of the equipment installed nearby is very important. Thus, the advice of using the same models suggested for frequency scans
[2] The bonding type changes the magnitude of the impedance at a resonance point. Thus, the simulations should be redone with a proper modelling of the major-sections if a resonance frequency is hit
[3] Except for high frequencies (see Sect. 3.5)

5.5 Systematic Method for the Simulation of Switching Transients

Several phenomena have been demonstrated through the several chapters of this book. At the same time, methods have been presented that can be used when studying different electromagnetic transients.

This section shows a schematic method for the study of electromagnetic transients in cables or for the realisation of an insulation co-ordination study.

The method hereby proposed is a generic method and it has to be adapted for the system that is being studied. Several of the steps are not necessary for many network configurations, whereas more detailed models/studies are required on others. Thus, it is strongly advised not to blindly follow the method described next, but to use the acquired knowledge to predict the more hazardous cases.

The method is divided into several phases, going from cable design and validation to the simulation of specific phenomena. It is also indicated where in the book can be found a more detailed explanation of each step.

Phase 1–Design cable
The first phase consists of the design and validation of one or more cables.

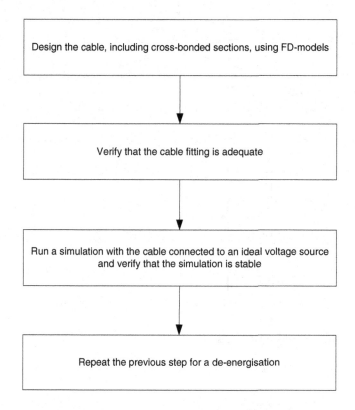

A Matlab code is available online and it can be used to verify that the model is properly designed and that the fitting is correct. The output of the code is the series and admittance matrices of a cable, as well as the respective frequency spectrum. These results can be compared with those obtained when using an EMTP-type software.

Phase 2–Simulations (resonance)
The second phase is to design the network and verify the existence of possible resonances.

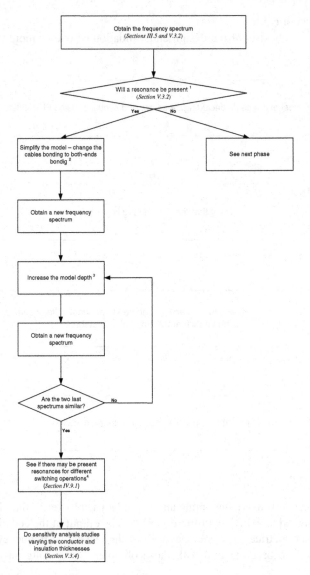

[1] By resonance, it is understood either the excitation of a harmonic in steady-state or a cable-transformer resonance

[2] The change depends on the frequencies of interest and the length of the cables. A good first test is to use the Matlab code available online to simulate the spectrum of the longest cable in the system and verify that the lower resonance point is inside the area of interest. If it is, the cable should not be simplified. This change is made with the sole purpose of enhancing the total simulation time

[3] The area modelled by means of lumped-parameters should be increased for resonance studies.

[4] The simplest of the two models is used

Phase 3–Energisation and de-energisation simulations

The third phase consists in the simulation of different switching operations to verify that the voltage and current limits are not exceeded.

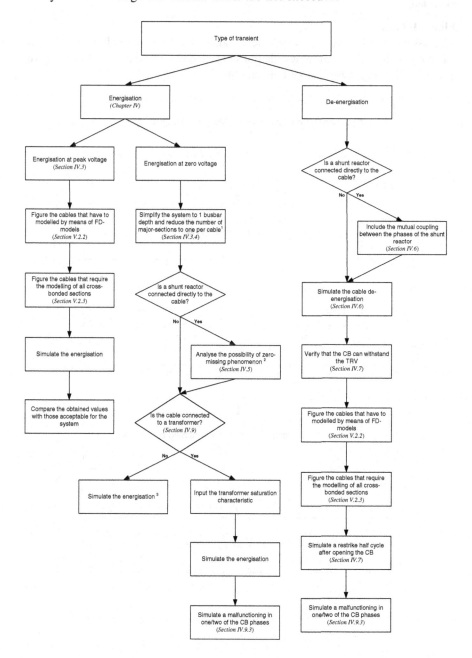

[1] This step is not necessary, but it will decrease the total simulation time without loss of accuracy. It is to be done if it is necessary to repeat the simulations several times.

[2] A simple first verification consists of seeing if the shunt reactor(s) compensate for more than 50 % of the reactive power generated by the cable.

[3] In these circumstances the presence of any undesired phenomena is not expected.

Phase 4–Faults

The last part is to study how short-circuits affect the waveforms. Remember, that this book focuses on electromagnetic transients and that this analysis does not give any indication regarding the system stability.

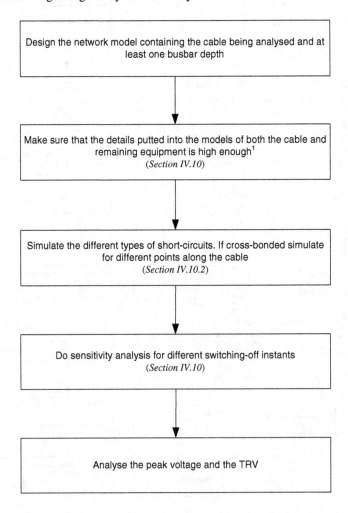

[1] As an example, the model of a transformer needs to include the stray capacitances to the ground as explained in Sect. 4.10.4.

5.5.1 Demonstration

To finalise the book we will try to solve an example for a cable of the West Denmark 150 kV network and observe several transient phenomena. The cable in question is installed between the nodes KAG and HNB, with a total length of 20.49 km. Figure 5.14 shows the single-line diagram of the system, including the adjacent cables that are required for an accurate simulation of the cable energi-sation/re-energisation.

In order to show different phenomena, the network will be slightly modified in some parts of the demonstration.

Phase 1

1. The cable and adjacent cables are designed using the datasheet and there are done the required corrections of the resistivity and permittivity (Sect. 1.1);
2. The fitting of all the cables is adequate and the simulation is stable. The time that passes between the energisation of the cable and the maximum peak voltage is 360 µs, corresponding to a coaxial mode speed of 170 m/µs for the KAG-HNB cable;

The cable modelling is adequate and it can be used in more complex simulations.

Phase 2

1. Using the file containing the entire network, it is estimated the number of lines that have to be included in the model during the simulation of the line ener-gisation. The method shows that besides the lines being energised, three other lines need to be included in the model (Fig. 5.14);
2. The method shows also that it is necessary a precise modelling of the bonding sections for all the cables. To be precise, it would not be necessary to design all bonding sections of the 40 km cable between FER and THO, as the major-

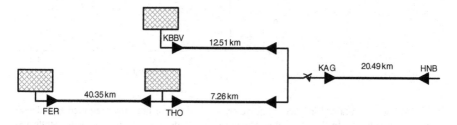

Fig. 5.14 Single-line diagram for a distance up to two nodes from the KAG node

sections closer to the FER busbar do not affect the waveforms, but for simplicity we do it;
3. The model is designed;

Figure 5.15 shows the voltage waveform during the energisation transient for four different modelling approaches:

- The system is under-modelled and does not include the FER-THO line (Model 1);
- Only the required nodes are included in the model (Model 2);
- The system is over-modelled and it includes more nodes than required (Model 3);
- Equivalent to model 2, the optimised model, but with the cables adjacent to the KAG-HNB cable bonded at both ends instead of cross-bonded. The cable being energised remains cross-bonded (Model 4);

The following results can be drawn from the simulations:

- The peak instant is the same for all four models;
- The magnitude of the peak voltage is the same for optimised model and the over detailed model. These two modes only start to diverge after the peak, at approximately 0.75s;
- The influence of the bonding is clearly seen in Model 4, moreover, this is the first model to diverge from the remaining models. However, since this model still has the depth required for the proper modelling of the coaxial mode waves, it has a peak overvoltage larger than Model 1.
- For some of the models the peak overvoltage is after the expected instant. This is due to an inaccuracy of the model as explained in Sect. 5.2.4. To avoid this problem, the modelling depth will increase, being the new lines modelled by means of lumped-parameter models.[17]

Fig. 5.15 Energisation of the cable for three different models. *Dashed line* Model 1; *Solid line* Model 2; *Dotted line* Model 3; *Dashed-dotted line* Model 4

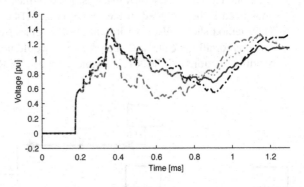

---

[17] As previously explained in Sect. 5.2.5 we should have more lines than required in order to avoid this inaccuracy. In this case, this was not done on purpose in order to demonstrate the problem.

Fig. 5.16 Frequency spectrum of the system seen from the receiving end of the cable. *Dashed line* Depth: 1 busbar; *dashed-dotted line* 2 busbars; *Dotted line* 3 busbars; *Solid line* 2 busbars and the remaining system modelled using lumped-parameters models

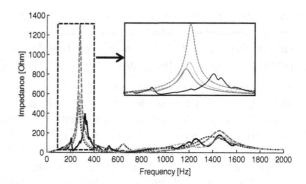

Phase 3

Figure 5.16 shows the frequency spectrum seen from the receiving end of the FER-THO cable, for different models. The figure shows that the more detailed the model the lower the magnitude at the resonance point, as explained in Sect. 5.3.2, and indicates the frequency of the transient for the different modelling approaches.

The spectrums also show that there is no need of concern with resonances in this case, as the magnitude at the resonance frequency is rather low for the more realistic models.

Phase 4

In order to increase the number of possible phenomena, a 31 Mvar shunt reactor compensating 80 % of the reactive power generated by the cable is installed. The shunt reactor is energised together with the cable and zero-missing phenomenon may occur.

Figure 5.17 shows that zero-missing phenomenon is present when the cable is energised at zero voltage. The cable is neither near a transformer nor connected in a weak point of the grid. Thus, no other problems are present.

Figure 5.17 compares also the waveforms for the detailed model and a simple model where the cable is connected to an ideal voltage source and bonded at both ends instead of cross-bonded. The comparison shows that for this specific

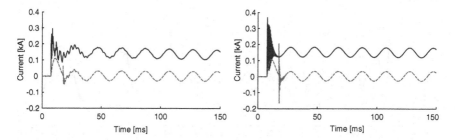

Fig. 5.17 Current through the CB during the energisation of the cable. **a** Detailed model; **b** Simplified model. *Solid line* Without pre-insertion resistor; *Dashed line* With pre-insertion resistor

phenomenon we can use simple models, without loss of accuracy, except during the first milliseconds, which are not relevant for zero-missing phenomenon. We can simplify the modelling even more by using Bergeron models or even lumped-parameter models and the results would still be accurate.

Energisation of a cable-transformer system

Before going into the next stage and simulate the de-energisation of the system, we will consider that a transformer is connected to the cable and see how the transient waveforms are affected. We will say that a 165/165 kV transformer is connected at the sending end of the cable in a first moment and at the receiving end of the cable in a second moment, i.e. we will simulate possible series and parallel resonances. We will simulate both phenomena for different modelling details. The models under analysis are:

- Model 1: The entire network is modelled by means of lumped-parameters models and the area around the point of interest is modelled using FD-models. The cables modelled by means of FD-models are those that where previous included in the model used to simulate the energisation;
- Model 2: The cables connected to the node of interest are modelled by means of FD-models and an N-port equivalent network is used;
- Model 3: Only the cable and the transformer are included. The rest of network is modelled by means of a Thévenin equivalent;

Figures 5.18 and 5.19 show the current and voltage in one of the phases during the energisation of the cable-transformer system for the three models. The accurate simulation of both phenomena needs a proper representation of the frequency spectrum at the node of interest, which requires the modelling of a large area of the system, but not the use of FD-models, as we have seen in Sect. 5.3.2. Normally, this is done in phase 3, when a resonance is seen in the frequency spectrum

Both figures confirm the theory as large differences are seen when comparing the reference waveform (solid line) with the other two waveforms.

De-energisation

We consider a normal de-energisation of the cable and shunt reactor together, as the restrike would be similar to a normal energisation, but with statistical

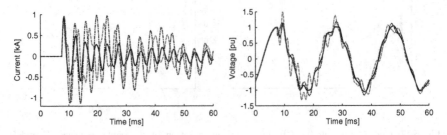

Fig. 5.18 Waveforms during the energisation-series resonance. **a** Current; **b** Voltage; *Solid line* Model 1; *Dashed line* Model 2; *Dashed-dotted line* Model 3

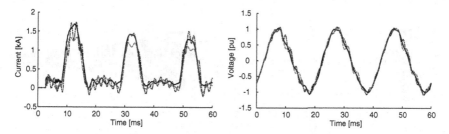

Fig. 5.19 Waveforms during the energisation-parallel resonance. **a** Current; **b** Voltage; *Solid line* Model 1; *Dashed line* Model 2; *Dashed-dotted line* Model 3

switching in order to find the worst-case scenario. Our objective is to see how the waveforms are during the de-energisation and if there is an overvoltage because of the mutual coupling between the phases of the shunt reactor. We know that for this phenomenon we only need to model the cable and the shunt reactor, which allows increasing the simulation speed.

However, before simulating a de-energisation it is necessary to verify the fitting parameters of the cable(s). It is not so unusual to see the voltage and the current raising to infinite after the switch-off of the CB. This instability is the result of fitting errors in the cable model. One way of verifying the accuracy of the model is to plot the eigenvalues of the model's Hermitian matrix for all frequencies and verifying that there are no negative eigenvalues, something that is far from being trivial. Another way of solving the problem, it is to simply use the trial-error method and change the fitting parameters of the cable and the time-step until the inaccuracy disappears.

This is not a big issue as we can easily see that something is wrong with the simulation, as both the voltage and current increase exponentially. However, in some situations the waveforms are still inaccurate, when simulating the de-energisation of a cable and shunt reactor together. The problem is that they do not increase to infinite, instead, they are rather similar to waveforms that can happen in other situations, something that makes detection of the inaccuracy rather complicate.

Figure 5.20 shows an example of this inaccuracy. In both cases it is simulate the de-energisation of the same cable and shunt reactor together, being the only difference on the fitting parameters of the cable.[18] Figure 5.20a is inaccurate, whereas Fig. 5.20b is accurate. We know that the first example is inaccurate, because a damping that fast would require a low impedance path to the ground, which is not present in the model. However, this may be challenging for less experience engineers, whom tend to be trustier on simulations. Another situation that is sometimes observed is the waveforms oscillating like if there was mutual coupling between phases, when such coupling does not exist.

---

[18] The cable length and fitting parameters were changed in order to show the inaccuracy. Thus, the cable used to simulate the waveforms shown in Fig. 5.20 is not the same used in the other simulations including Fig. 5.21.

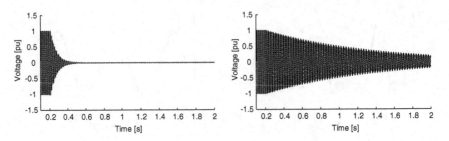

Fig. 5.20 Voltage at the sending end of the cable during the de-energisation. **a** Inaccurate simulation; Accurate simulation

Fig. 5.21 Voltage at the sending end of the cable during the de-energisation considering mutual coupling between the phases of the shunt reactor ($M_{AB} = -0.05H$; $M_{BC} = -0.05H$; $M_{AC} = -0.03H$)

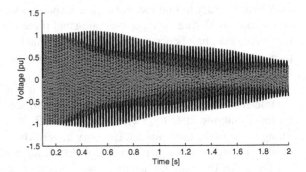

A way to deal with this problem is to simulate the de-energisation for different fitting parameters and see if the results stand. However, the most important is to be always critic of the results and verify that they are according to the expected.

We are now ready to simulate the de-energisation of the cable with mutual coupling between the phases of the shunt reactor. Figure 5.21 shows the voltage at the sending end of the cable during the de-energisation and it is possible to observe an overvoltage in one of the phases as well as the existence of more than one frequency.

Figure 5.22 shows the voltage and current waveforms during de-energisation of the cable for a faulted switch-off of the CB. The simulation is done with and without a transformer in order to see if there will be ferroresonance. The simulations show that for this specific case, ferroresonance is not of concern.

Phase 4

For last, but not least, we simulate short-circuits in different points of the cable. There are simulated single-phase-to-ground, two-phases-to-ground and three-phases-to-ground. The simulations are typically made considering the fault at two different points of the cable and for different CB disconnection sequences.

Figure 5.23 shows only the single-phase-to-ground and the two-phases-to-ground fault, with and without shunt reactor, and considering that the CB of the KAG side opens last; the remaining cases can be done by downloading the PSCAD simulation file available online.

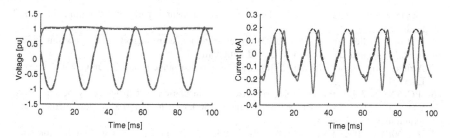

Fig. 5.22 Voltage and current waveforms during the de-energisation when one phase of the CB does not open. **a** *Solid line* Voltage with transformer; *Dashed line* Voltage without transformer; **b** *Solid line* Current with transformer; *Dashed line* Current without transformer

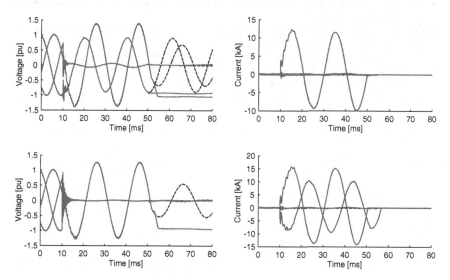

Fig. 5.23 Voltage (left) and current (right) during a SPGF (up) and a TPGF (down). *Solid line* Without shunt reactor; *Dashed line* With shunt reactor

For this particular case, the waveforms present a normal behaviour and there are not required extra precautions.

References and Further Reading

1. Morched AS, Brandwajn V (1983) Transmission network equivalents for electromagnetic transients studies. IEEE Transactions Power Apparatus Syst 102(9):2984–2994 September 1983
2. Wiechowski W, Børre Eriksen P (2008) Selected studies on offshore wind farm cable connections—challenges and experience of the danish TSO. In: Conference on IEEE-PES General Meeting, Pittsburgh

3. Martinez-Velasco Juan A (2010) Power system transients—parameter determination. CRC Press, Boca Raton
4. Watson Neville, Arrillaga Jos (2003) Power systems electromagnetic transients simulation. IEEE Power and Energy Series, United Kingdom
5. Arrillaga Jos, Watson Neville (2001) Power system harmonics, 2nd edn. John Wiley & Sons, England
6. IEC 62067 (2004) Power cables with extruded insulation and their accessories for rated voltages above 30 kV (Um = 36 kV) up to 150 kV (Um = 170 kV)—test methods and requirements, 3rd edition
7. IEC 60840 (2001) Power cables with extruded insulation and their accessories for rated voltages above 150 kV (Um = 170 kV) up to 500 kV (Um = 550 kV)—test methods and requirements, 1st edition
8. IEC TR 60071-4 (2004) Insulation co-ordination–Part 4: Computational guide of insulation co-ordination and modelling of electrical networks
9. Cigre Joint Working Group 21/33 (2001) Insulation co-ordination for HV AC underground cable system. Cigre, Paris
10. Cigre Working Group C4–502 (2013) Power system technical performance issues related to the application of long HVAC cables. Cigre, Paris
11. Cigre Brochure 39 (1990) Guidelines for Representation of Network Elements when Calculating Transients, Working Group 02 (Internal overvoltages) Of Study Committee 33 (Overvoltages and Insulation Coordination), Cigre, Paris

Erratum to: Electromagnetic Transients in Power Cables

Filipe Faria da Silva and Claus Leth Bak

Erratum to:
F. F. da Silva and C. L. Bak, *Electromagnetic Transients*
in Power Cables, **Power Systems,**
DOI 10.1007/978-1-4471-5236-1

This book should be published with extra server material. It was missed out in published volume. Now it has been added.

The online version of the original book can be found under DOI 10.1007/978-1-4471-5236-1

F. F. da Silva (✉)
Department of Energy Technology, Aalborg University, Aalborg, Denmark
e-mail: ffs@et.aau.dk

C. L. Bak
Institute of Energy Technology, Aalborg University, Aalborg, Denmark
e-mail: clb@et.aau.dk

F. F. da Silva and C. L. Bak, *Electromagnetic Transients in Power Cables*,
Power Systems, DOI: 10.1007/978-1-4471-5236-1_6,
© Springer-Verlag London 2013

Printed in the United States
By Bookmasters